Ecoviews

Ecoviews

Snakes, Snails,

and Environmental Tales

Whit Gibbons
Anne R. Gibbons

Foreword by John Cairns, Jr.

The University of Alabama Press
Tuscaloosa and London

∞
The paper on which this book is printed meets the minimum requirements of American National Standard for Information Science-Permanence of Paper for Printed Library Materials, ANSI Z39.48-1984.

Library of Congress Cataloging-in-Publication Data

Gibbons, Whit, 1939–
Ecoviews : snakes, snails, and environmental tales / Whit Gibbons, Anne R. Gibbons ; foreword by John Cairns, Jr.
p. cm.
Includes index.
ISBN 0-8173-0919-5 (pbk. : alk. paper)
1. Natural history. 2. Ecology. 3. Human ecology. I. Gibbons, Anne R., 1947– . II. Title.
QH45.2.G335 1998
508—dc21 97-37826

British Library Cataloguing-in-Publication data available

To Bill Fitts, for all he does,

and in memory of our mother

A land ethic for tomorrow should . . . stress the oneness of our resources and the live-and-help-live logic of the great chain of life. If, in our haste to "progress," the economics of ecology are disregarded by citizens and policy makers alike, the result will be an ugly America.

—Stewart Lee Udall, *The Quiet Crisis*

Contents

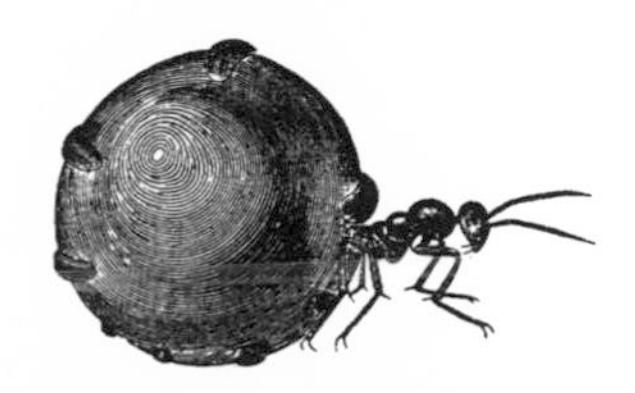

Foreword

The day I received the manuscript for this book from Whit Gibbons, I also visited his laboratory at the Savannah River Ecology Laboratory where animals are kept for visiting schoolchildren to see and touch. Only a few minutes in that laboratory made me abundantly aware of the tender, loving care given these species that would link schoolchildren to the natural world in a variety of ways. The animals themselves were bursting with health, and their cages were clean, with appropriate habitat replication—the kind of care one would expect from people who have a strong relationship with other species. The evening before, I had an opportunity to talk with Whit's sister Anne about the ways in which an esteem for natural systems and the species with which we share this planet could be communicated to everyone, including schoolchildren and even preschoolers. Clearly, the Gibbons siblings are actively caring people, and this book shows it.

In this era of emphasis on various "rights," we sometimes forget that we have intruded into almost every natural system on Earth. This intrusion inevitably involves risks, and the authors discuss some of those risks to human beings, from the stinging bite of a centipede to the more lethal bite of a crocodile. More to the point, however, the authors devote considerable discussion of species' vulnerability to us and suggest that we need to learn to share our environment with tolerance toward both nature and wild creatures.

The book is refreshingly free of academic jargon. The authors have accomplished this feat without sacrificing good science, as is, unfortunately, all too common in the news media these days. In *Ecoviews,* they make the telling point that our environmental attention span is too short, contrasting it to the public's (and the media's) daily fascination with sports results. All forms of mass media record daily and weekly victories and defeats throughout a sport's season, with details often no more burdensome to learn than those needed for becoming moderately environmentally literate. People can become better informed about nature, becoming fans of the environment as well as of sports.

The book is replete with information that has escaped even an environmental junkie like me—information, for example, on Colorado River toad smoking. Anyone who is interested in the environment or anyone who enjoys association with natural systems will relish reading about the personal experiences that fill this book. The examples are not just nature activities, such as bird watching, that traditionally seem to attract the most public attention. They include also such seemingly common experiences as teaching a child to smell flowers and to wonder why their perfumes are different. The authors suggest that we teach ourselves to view everyday natural occurrences with ecological awareness. They provide an excellent series of illustrative examples that could easily fulfill environmental teaching needs virtually anywhere in the world.

When this book is published, I intend to get copies for all five of my grandchildren and their parents and for the children of colleagues in the laboratory here. I am always delighted to see books that celebrate our relationship with natural systems, written by people who not only have a good grounding in science but also revel in their interactions with other species. I hope this book receives the wide attention it deserves.

JOHN CAIRNS, JR.
University Distinguished Professor
of Environmental Biology Emeritus
Virginia Polytechnic Institute
and State University

Preface

These days, one would have to be out in what is left of the ozone not to be at least a little bit concerned about endangered life on Earth and our vanishing natural environments. The planet itself does not need to be saved; it will continue to circle the sun. In fact, life on earth will continue in some form no matter how human beings treat the environment. But human life itself is a different matter. The issue at stake is whether people are capable of modifying their situation and reversing what some view as "the current trend toward self-destruction."

We think we are. But to do so, people, collectively and individually, must accept the intrinsic worth of all the organisms that dwell on the third planet from the sun. They must recognize that a unified effort to preserve the Earth's biodiversity will benefit all of us. They must establish—locally, regionally, nationally, globally—an attitude more reminiscent of the Three Musketeers than of Genghis Khan. The world is not ours to conquer. Until we realize that, and act accordingly, we are in danger of destroying the very components that help support our life on Earth.

Maintaining the integrity of our planet's biodiversity is critical for our survival. The passage of the Endangered Species Act in 1973 represented a major step toward protecting species and offered a powerful national statement about the sanctity of species other than our own. But most ecologists and many other people know that preserving focal species is not the real goal. We need to protect biodiversity across a broad spectrum of habitats. The Endangered Ecosystem Act is called for now. Legislation alone will not protect species or their habitats. Such legislation must be coupled with the attitude that the laws are proper, the goal is worthwhile, and the benefits outweigh the costs. Our intent in this book is to encourage that attitude, to promote the idea that we are all in this together: all people, all plants, all animals, and all their natural habitats.

Familiarity is a potent tool for engendering support for a beleaguered group, whether it be humans, other animals, or plants. Becoming familiar with the traits, lifestyles, and associations of a group different from our own helps us to be more compassionate, more tolerant, more sensitive to that group's needs. Familiarity with another culture breeds respect and understanding, not contempt. Each plant and animal species on the planet represents a different culture from our own, no matter what human subculture we are associated with. The more we learn about these cultures the more likely we are to accept their inherent value.

We tend to appreciate others who have qualities and characteristics we esteem or would like to possess ourselves, whether such traits manifest themselves in everyday behavior or in extraordinary fortitude. Plants and animals meet these criteria, although the general public may be unaware of the connections. Animals from aardvarks to zebras display instinctive behaviors that qualify for what some people think of as "family values." For example, a mother alligator guards the nest that holds her incubating eggs and becomes fiercely protective of the babies for months after they hatch. Honeybees will give their individual lives to protect the hive and colony.

Other animals can justly be called phenomenal. The silver ants of the Sahara maintain a delicate balance between becoming prey of a desert lizard and surviving the intolerable temperatures of the desert sun. And virtually all ants are stronger for their size than Samson or Hercules. For those who admire high ideals and strong work ethics, a cast of thousands can be found in other species.

In addition to appreciating admirable behavior, most people have compassion for the downtrodden. Anyone who thrives on supporting the underdog has abundant opportunity for doing so among the plants, animals, and natural habitats of Earth. Humans are the world's environmental bullies, and almost every other species on Earth can be considered oppressed and persecuted, albeit sometimes unwittingly.

Our goal in this book is to engender esteem for the wealth of biodiversity on Earth. One definition of *esteem* is "to regard highly and prize accordingly." If we are, indeed, to reverse the current trend toward self-destruction, we must make a unified effort to protect *all* species of the world and their habitats. Let us relinquish the attitude that the Earth and its inhabitants are ours for the taking. The time has come to forgo our self-appointed role as rulers of a self-proclaimed domain, to abandon the exaggerated notion of our own self-importance. Failure to do so will not destroy the planet we know as Earth, which will continue to spin on its axis. But without clean air, clean water, and reforestation, without the biodiversity inherent in this fragile ecosystem, *we* cannot endure. The issue is not whether we can save the planet; it is whether we can save ourselves and in such a manner that all who populate the Earth are content with their position in the world ecosystem.

To accomplish our stated objective—to engender esteem for the wealth of biodiversity on Earth—we offer perspectives about the ecology of plants, animals, and habitats that we found to be some combination of fascinating, thought provoking, and instructive. The studies we use as examples not only are technically accurate, constituting what scientists refer to as "good science," but also qualify as entertaining to nonscientists. We try to keep preaching to a minimum, but if an occasional soapbox creeps in before the last chapter or two, we apologize.

Authors' Note: Some of the vignettes that follow were originally published in newspaper columns in the first person singular, although both of us were involved in the writing and editing of the columns. We have preserved the use of "I" in those instances where it makes for a better tale or explanation or when a tale actively involves just Whit in the course of his fieldwork.

Because of the diversity of academic titles among the researchers we cite, we have dispensed with titles. The affiliations of researchers mentioned in the text are those at the time their research was conducted and do not necessarily reflect their current affiliation.

Acknowledgments

We are grateful to the many scientists who contributed to this work by reviewing what we had written about their research. We took most of the excellent suggestions that were made; any errors are our own. In particular, we thank Bill Amos (Cambridge University), Steven N. Austad (University of Idaho), Randall D. Babb (Arizona Game and Fish Department), H. Bernard Bechtel (dermatologist in private practice, Valdosta, Georgia), Stephen H. Bennett (South Carolina Department of Natural Resources), Vicki M. Boatwright (U.S. Fish and Wildlife Service), J. Russell Bodie (University of Missouri), David Boughton (University of Texas), David E. Brown (Arizona State University), Vincent Burke (University of Missouri), Janalee P. Caldwell (University of Oklahoma), Geoffrey M. Clarke (CSIRO Division of Entomology, Canberra, Australia), James P. Collins (Arizona State University), Steven D. Garber (Rutgers University), Jaap Graveland (Institute for Forestry and Nature Research, the Netherlands), Pam Graves (University of Kentucky), Joe Johnson (Kellogg Biological Station of Michigan State University), W. Joe Lewis (U.S. Fish and Wildlife Service-ARS, Coastal Plain Experiment Station, Tifton, Georgia), Harbin Li (USDA Forest Service), Gene E. Likens (Institute of Ecosystem Studies, Millbrook, New York), Jeff Lovich (U.S. Geological Survey, Biological Resources Division), Ken Marion (University of Alabama at Birmingham), William D. McCort (the Nature Conservancy), Anne Platt McGinn (Worldwatch Institute), Peter K. McGregor (Behaviour and Ecology Research Group, University of Nottingham), Michael Alexander Menzel (University of Georgia), Joseph C. Mitchell (University of Richmond), Robert R. Parmenter (University of New Mexico), Russell B. Rader (U.S. Fish and Wildlife Service, Rocky Mountain Station, Research Division), Curtis J. Richardson (Duke University Wetland Center), Robert E. Ricklefs (University of Missouri—St. Louis), Kenneth G. Ross (University of Georgia), Pietro Scotto (Department of Physiology, University of Naples, Italy), Paul Sherman (Cornell University), Rick Shine (University of Sydney), Peter Stangel (National Fish and Wildlife Foundation), Carl C. Trettin (USDA Forest Service), Ricardo Villalba (Laboratorio de Dendrocronologiá, CONICET, Argentina), Ridiger Wehner (University of Zurich), Sibylle Wehner (University of Zurich), G. W. Max Westby (University of Sheffield, England), Howard Whiteman (Murray State University).

We also thank students, technicians, and faculty from the University of Georgia's Savannah River Ecology Laboratory: I. Lehr Brisbin, Kurt A. Buhlmann, Sarah Collie, Justin D. Congdon, Adrienne DeBiase, Mike Dorcas, Nat B. Frazer, Judith L. Greene, Mark S. Mills, Tony Mills, Margaret E. Mulvey, Michael C. Newman, Joe Pechmann, David Scott, Barbara Taylor, Tracey D. Tuberville, S. Rebecca Yeomans. Thanks also to Gary Gore for his skillful and creative graphic design.

Special thanks go to Patricia J. West of Tuscaloosa, for reading and editing almost all the original newspaper columns before their submission. Her insights into the complex interaction between people, politics, and environments have added greatly to the interpretation and presentation of the material.

Ecoviews

1

Do Trees Own Themselves? Thoughts to Ponder

"We by-and-by discovered, however, what I thought well worth my trouble, a pair of those splendid birds, the Ivory-billed Woodpeckers. . . . They were engaged in rapping some tall dead pines, in a dense part of the forest, which rang with their loud notes. These were not at all like the loud laugh of the Pileated . . . , nor the cackle of the smaller species, but a single cry frequently repeated, like the clang of a trumpet. . . . We succeeded in shooting both, which I skinned and dissected."

The ivory-billed woodpecker is now presumed extinct in North America. The above account was written in southern Alabama in the late 1830s. No illegal act was committed and no environmental harm was intended by the writer, Philip Henry Gosse. His 1859 book, *Letters from Alabama,* reissued in 1993, contains verbal portraits and is an excellent addition to our knowledge of plants, animals, and natural habitats in the 1800s.

William Bartram is generally recognized as having written the most thorough accounts of natural history in the southern regions of the country during the early colonization period, in the late 1700s. More than half a century after Bartram, Gosse's views as a naturalist give intriguing insight into the natural history of the developing region. Gosse arrived in Mobile, Alabama, in 1838, and his writings disclose environmental, as well as social, attitudes of the times. An Englishman who had lived in Philadelphia, Gosse came to Alabama with prospects of being a teacher. His "letters" are notes and sketches based on his observations of the region, its people, and its wildlife.

Gosse mentions more than three hundred plants and animals (usually including scientific names). Most are insects: he had a special fondness for butterflies and moths. He also reports observations of about fifty birds, a dozen reptiles, thirty-five trees, and more than fifty flowers, shrubs, and vines. His accounts of natural history reveal a keen and observant individual with a background in biology. An annotated section by

Daniel D. Jones and Ken R. Marion at the end of the modern edition is valuable for today's readers; it cites currently accepted names of each species of plant or animal mentioned by Gosse.

Even without the annotation, many of Gosse's descriptions are accurate enough for readers to recognize today's extant species and some that are no longer with us. For example, Gosse states that on "a ride to Selma . . . I had the pleasure of seeing a flock of Parrots. . . . There were eighty or a hundred in one compact flock, and as they swept past me, screaming as they went, I fancied that they looked like an immense shawl of green satin, on which an irregular pattern was worked in scarlet and gold and azure." These were of course now-extinct Carolina parakeets. The last confirmed specimen of the Carolina parakeet died in the Cincinnati Zoo on 21 February 1918. Ironically, the last passenger pigeon had died in a nearby cage in the same zoo four years earlier.

Some readers will be alarmed at the prevailing attitude of the times that any living animal was fair game for a man with a rifle. Gosse, like John James Audubon, shot birds in order to study the specimens. But he mentions the indiscriminate shooting of other animals by anyone holding a gun. The prey included great horned owls, Mississippi kites, and opossums—not for food but because of an attitude that shooting anything was all right. Even Gosse, clearly a lover of nature, is not judgmental on the topic. This attitude is understandable. As recently as the 1950s shooting even songbirds or hawks went unchallenged in many parts of the country. Unfortunately, this same attitude persists in some regions today, although a more environmentally aware society is gradually pressing such views into extinction.

Overall, Gosse's book is a reflective and informative guide

for those interested in the perspectives of a biologist in the South a century and a half ago. But environmental issues have become increasingly complex not only within the last century but also within the last few decades. In the 1970s, around the time of the first Earth Day, most people could easily decide which position they should take in an environmental debate. Each year, the decisions have become less and less clear-cut. Economics, water quality, agricultural needs, disappearing forests, and other complex issues must be pondered, evaluated, and placed in a proper balance. The environmental conscience of most of us is taxed daily. The American beaver provides an outstanding example of the complexity of environmental conscience at the personal level. The perplexing problem of the beaver mirrors, in microcosm, the ongoing environmental dilemmas that we face collectively and individually.

Enter the Beaver

As many as 400 million beavers inhabited North America in the 1700s. By the early 1900s fur trappers and disgruntled landowners had reduced the beaver population to near extinction in most of its range. Today, beavers have made a comeback and can be found in many regions where they were rare or absent for decades.

Beavers are undeniably cute, from the chubby babies dragging little paddlelike tails behind them to ones that have been raised as pets to adulthood and can even be fed safely by hand. A family of beavers in a lake can be the pinnacle of fascination for someone who enjoys observing animal behavior—a wildlife experience with a lesson in sociology. If they do not feel threatened, beavers will busily swim about building a lodge or dam as a cooperative family unit, although they do not build either lodges or dams in some situations. Contented beaver family members can even be heard mewing to one another. It is easy to be smitten with their industrious, friendly behavior. Their hard-working nature is a textbook example of a positive trait we would all do well to acquire.

People who live on a lake, however, may find that a family of beaver neighbors can soon present a dilemma. Having their own private beaver family involves both costs and benefits. For example, the residents along a lake spoke with pride of their small beaver colony. But one spring night the beavers cut down and carried away six boxwood shrubs planted the previous afternoon. The next night, as if to make a firm statement to their previously admiring audience, the beavers expertly removed a flowering pink dogwood tree that grew only a few feet away from the boxwood stumps. Because the record-size tree (a cottonwood in Canada) known to have been felled by beavers was five and a half feet in diameter, a landowner could easily become concerned. The predicament is how to keep beavers for show-and-tell and yet have them behave to our liking. Pam Graves, an environmental educator who has kept a pet beaver for years, has suggested that if people want to keep beavers in their ponds, they should plant red maples and willows. The beaver teeth will act like pruning shears, but the trees will not be killed.

Another common problem occurs when beavers decide the water level of a lake is not high enough to suit them. They may proceed to dam up the overflow pipe, thus flooding driveways and backyards. Removing the vegetation from the pipe will help for only a day at a time, because beavers, some of whom do not care much for running water, will repair a broken dam or fill an unclogged pipe within hours. Beavers also burrow under roads and weaken structural foundations. Another stressful situation can arise when beavers build a lodge inside a boathouse. Although the activity may be fun to watch, the attractiveness begins to wane when the beavers use the boathouse pilings to increase the size of their lodge. In each case, the dilemma is, do you get rid of the beavers or forfeit your personal property? (Beavers, however, are not the only species that can damage property with dams. As Pam Graves points out, beavers in the Savannah River that separates South Carolina and Georgia build neither lodges nor dams but instead live in holes they make in the bank. She notes that the only dams on the Savannah River have been put there by human activity. Think how much habitat, including formerly private property, these dams have flooded.)

The quandary with beavers is compounded when people grow fond of the flat-tailed animals around them and do not want to hurt them. The simplest way to remove beavers is with a steel snap trap that kills them. This method does not qualify as "not hurting." So what about trapping them alive? With more effort than required for snap traps—and less effective results—beavers can be captured unharmed. Then the question arises, what do you do with them? Regional zoos have a limited demand for beavers, so this option is soon exhausted. Releasing what you consider a pest into another lake or stream hardly seems fair to whoever lives there. Besides, like many animals, beavers will return to their former home if the distance is not too great. So, what do you do?

No simple answer exists. Beavers put the issue of nuisance wildlife on a personal scale. You have two clear options: kill the beavers or ignore the destruction of your property. A compromise is often out of the question.

The issue of balancing one's love of wildlife with decisions about personal lifestyle, including ownership of pets, can be even more complicated. One dilemma that affects millions of Americans is caused by a predator that lurks in virtually all terrestrial habitats, ready to pounce upon any small prey—mammal, bird, or reptile. The creature is everywhere, from England to Australia, and throughout North America. Biologists consider it the most dangerous carnivore in many regions because of its large numbers, stealth, and agility. An inclination to hunt, whether hungry or not, makes the species a potential menace to all wildlife. The killer is the domestic cat, a species introduced to North America in the 1600s, centuries before fire ants or kudzu.

Look What the Cat Dragged In

An article in *Virginia Wildlife* by Joe Mitchell of the University of Richmond gives some striking facts about house cats and their potential impact on native wildlife. He distinguishes between "domestic, free-ranging cats" (those that spend much of their time outdoors but are assured of a food supply at home) and feral (wild) cats. The latter have no human home and therefore must provide their own meals. Both types of cats prey on wildlife and are highly successful predators.

Mitchell, a biologist who lives in a suburban neighborhood and who is a cat owner, kept a tally of the wildlife trophies his family's four cats brought home over a period of eleven months. The total was 104 individuals of 21 native species: 6 kinds of birds, 8 kinds of mammals, 7 kinds of reptiles. Among the prey were flying squirrels, chipmunks, Carolina wrens, and cardinals. Peter Stangel, with the National Fish and Wildlife Foundation in Washington, D.C., told of his two cats' kill records during the time he lived in rural South Carolina: 15 different species of birds, mammals, and reptiles in four months. Anyone with a cat knows such prey numbers are not unusual for active domestic cats with access to wooded areas. Such tallies, which include only animals brought to the homeowners, are probably underestimates. On the other hand, some cats do not seem particularly predatory. Some fat, lazy felines lounge and bask all day and sleep inside at night.

Mitchell used his tally to provide some measure of the total impact that house cats might have on local wildlife. The Humane Society estimated about a million house cats in Virginia, not counting the feral ones. Mitchell calculated that 3 million songbirds and 27 million native mammals potentially fall victim to domestic cats annually, if all kill at the rate that his did. This record is only for one year and only in Virginia! An article by Paul Karr in *Sanctuary*, the journal of the Massachusetts Audubon Society, provided comparable data on wildlife destruction by domestic cats. An estimated 2 million birds are killed annually in that state. The article mentions another study stating that no fewer than 20 million birds are killed by cats each year in Britain. Cats are also considered a major menace to wildlife in some parts of Australia, where active eradication programs have been instituted by some of its citizens.

Awareness that cats can have a major impact on wildlife is not new. More than a century ago in England, cats were recognized as affecting plant, as well as animal, communities. Red clover was considered dependent on "humble bees" for pollination. The number of humble bees and the level of successful pollination were low in some areas, where field mice destroyed the hives. However, when cats were present, the number of mice was generally low. The author reporting this observation concluded, "it is quite credible that the presence of a feline animal in large numbers in a district might determine . . . the frequency of certain flowers in that district!" The author was Charles Darwin; the book was *The Origin of Species.*

Human beings obviously appreciate cats, as evidenced by the presence of an estimated 60 million in the United States. Their impact on small native animals is significant and in many instances disturbing, whether one is concerned about disappearing wildlife or concerned that the gentle house cat could get a bad reputation. Will Americans concerned about the national decline in numbers of songbirds and other wildlife be willing to get rid of their cats? Not likely. So what should we do about an environmental problem of this kind?

Can we rationalize that cats are only filling a void left by the disappearance or decline of natural predators like panthers, bobcats, and wolves? Do we refer to Darwin's observation that cats can be important in the preservation of flowers? No matter what justification we offer for not getting rid of our cats, such as emphasizing that they do not destroy natural habitat the way some other species do, the reality is that they kill a lot of small wildlife. One solution for the situation with house cats is simply to keep them indoors, a position strongly recommended by many as best for the cat as well as the wildlife.

Cats are not the only nonnative species introduced into a habitat where it competes with native species. Imagine the scene of a herd of horses feeding on an open range. A majestic stallion watches over his harem of mares on a green plain that stretches from one blue horizon to the other. Environmental beauty? Or environmental destruction?

Who's at Home on the Range?

In looking at a magnificent thoroughbred horse, we sometimes forget that the ancestors of all pets and domestic animals were once wild. Each adapted to a specific natural environment before coming under the care and protection of people. The process often goes full circle, as with the feral horses that now roam parts of the West. Horses, like cats, are still capable of living in natural environments, even after generations of care and breeding by humans. The descendants of formerly domestic equine stock, after many centuries of complete or partial domestication, can clearly live under sometimes harsh environmental conditions. Good examples are the ponies of Chincoteague and Assateague Islands, the Atlantic barrier islands off the Virginia coastline. The ponies' ancestors escaped from wrecked European sailing vessels. The descendants have survived now for many generations in the sometimes stringent environments of the coastal islands.

Owners of thoroughbred horses are proud if they can identify their steeds' ancestors back to the turn of the century. Thoroughbreds have long genealogies, recorded in careful detail. But the search back through history can only go so far, for experts disagree about the origins of the domestic horse; presumably its roots were in a grassland region of Europe or Asia. The earliest records of human beings riding horses are from Asia, fifty centuries ago, near present-day Iran.

Paleontologists probably smile at such short-term records. The lineage of horses with which paleontologists work can be traced back 65 million years. We know from fossils that the ancestral "dawn horses" (*Eohippus*) were about five hands high (less than two feet) and lived in North America and Europe. Four flat toes allowed them to walk in a swampy environment; their short teeth were suitable for a habitat filled with lush, leafy vegetation. As centuries passed, their environment changed to a firmer terrain with coarse grasses. This new diet wore down short teeth very quickly. Thus, horse ancestors with longer teeth were favored for survival. By following fossil remains through geologic time, we see the horse developing the single hoof, as we know it today. Also, as horses spread throughout all continents except Australia and Antarctica, they became larger, owing to unknown forces of natural selection.

According to William D. McCort of the Nature Conservancy, herds of wild horses ranged across North America as recently as eleven thousand years ago. Then, horses disappeared from the Western Hemisphere. Although the return of domestic animals to the wild is seldom good for the overall environment, one might claim that horses in the New West are former natives that were simply absent for a time—a long time.

How long does an introduced species have to be present before we call the situation "natural"? Should horses that have now been around for centuries be considered part of the natural environment? And who should make the decision about whether a habitat should be returned to its original state or left to take its own course? These questions have no simple answers—they involve not only ecology but also politics and economics. Complicating the issue is the fact that cattle are allowed to graze unhindered on many open ranges of the western United States. Cows cause more damage to native vegetation than do horses. Should we be overly concerned for the environmental impact caused by horses when we allow cattle, which are estimated to outnumber the horses by more than four hundred to one, to have free rein over thousands of square miles of public lands?

This is not an easy issue to resolve. Who has the greater rights? The horses, because we feel a closeness to them and because they have clearly found a suitable home? Cows, because we eat them? Or the natural plants and animals of North America that have, for more than ten thousand years, lived and evolved into an existence with neither horses nor cows? Or do we quit trying to referee the situation and let whoever happens to have arrived sort the problem out among themselves?

The Palm That Keeps Its Feet Wet

Anyone familiar with kudzu in the southeastern United States or autumn olive in the Midwest can readily appreciate that environmental quandaries created by nonnative, introduced species are not restricted to animals. On a trip to Palm Springs, California, I learned of a plan to bring convicts into the desert to apply chemicals that would poison a particular type of shrub, an introduced species that threatened the environmental integrity of the desert ecosystem. I was in a jeep with Jeff Lovich and Roland de Gouvenain, who were at that time officials with the Department of Interior's Bureau of Land Management (BLM). We were driving to a place called Dos Palmas, a unique desert area with many native California fan palms.

BLM's mission is to manage public lands and resources in a manner that best serves the needs of the American people. The lands administered by BLM in California make up more than 17 million acres, an area larger than the state of West Virginia. The resources include recreation, wildlife, cultural values, timber, minerals, and wilderness. Jeff and Roland are ecologists, so they focus on preserving and managing native plants and animals in sensitive environmental areas.

We passed many different kinds of palm trees on the way to Dos Palmas, but only one kind, the California fan palm, is native to the state. All the others are species introduced from other states or other countries. I had heard about "the palm tree that likes to keep its feet wet," a phenomenon known in the region since the earliest settlers arrived. Any thirsty desert traveler was ecstatic at the sight of palms; under natural conditions, palms indicated the presence of water, a true oasis. To see the palms today in their natural setting at one of the two major oases at Dos Palmas is like stepping into an ecological dreamworld.

We sat beside a clear pool, surrounded by towering palm trees that shield not only the sun but any sight or sound of the surrounding desert. The solitude of the oasis bespeaks protection and security. The trees are about fifty feet tall with trunks two feet across. The protection from sun and wind comes from the brown palm fronds that hang from top to bottom of each tree. The top fans of the palms are green, but as each season's fans are replaced by new growth, the dead ones droop toward the ground, adorning the tree like a huge skirt. The stand of palm trees creates a series of thick, heavy curtains on all sides, some more than fifteen feet in diameter. Desert completely surrounds the few acres of water and palm trees, which serve as refuges for desert birds, diamondback rattlers, the endangered desert pupfish, and a host of small mammals.

The desert has a beauty of its own. Creosote bushes and mesquite trees are scattered across the rocks and sand. On a warm winter day lizards scamper from one clump of vegetation to another. Natural sounds. Natural landscape. Native inhabitants to observe. This is how nature really is and should be.

But an invader spoils the scene at Dos Palmas. Amid the native desert plants, even around the thick stand of palms, a nonnative has emerged, and it threatens to change the scenery. A Eurasian shrub or small tree—the tamarisk, or saltcedar—has become established at Dos Palmas. Tamarisk is a tough competitor for native plants under certain conditions: it overshades other vegetation and uses great quantities of the precious soil moisture. A mature tamarisk tree produces half a million wind-blown seeds each year, and any area of moist disturbed soil can serve as a germination site for a tamarisk seed. Native animals make little use of the trees for food or shelter, and only a few birds nest in them.

BLM's directive was to return Dos Palmas to its natural state by eliminating tamarisks from two thousand acres. Simply cutting them down is not effective because they resprout vigorously. Even cut limbs can take root. Thus BLM has developed a tamarisk removal plan. The plants will be cut, and herbicides that have been carefully tested and judged environmentally safe will be applied directly to the cut stems. One plan was for the labor-intensive project to be carried out under BLM's supervision by prisoners from a state correctional facility.

Dos Palmas, with its admixture of palm oases, natural desert, and alien shrubs, suggests that southern California has the potential to qualify as environmental chaos. Some think it has already achieved this condition. Others, including some at BLM, believe many of the natural areas can still be reclaimed. Should we launch a desert attack on an imported weed with the aid of tree-poisoning convicts? Some would argue that a tree introduced more than a century ago should not be eliminated. But if we want a natural desert, the alien tree must go. The project is a major undertaking in environmental management. When dedicated ecologists with no political agenda are the environmental managers, why not give them a try?

Who Owns That Tree?

Trees, however, need not be introduced aliens to create controversy. A "man against tree" battle occurred in the 1990s in the Low Country of South Carolina where three big trees were declared to be standing in the way of education. Each tree was more than two feet in diameter and had occupied the same spot for eighty years. That spot was now needed for new classrooms. Simply put, a legal contest ensued over whether the school district was required to comply with a county ordinance that protected the trees because of their size. One must wonder if school officials pondered the point of whether next year's environmental education classes should be delivered in crowded classrooms shaded by big trees or in a vacant lot with a roomy building.

Another Low Country controversy, near Charleston, South Carolina, resulted from plans to cut down huge live oak trees along the road that leads through Johns Island to Kiawah Island. A major professional golf tournament, the Ryder Cup, was scheduled for the following year, and a decision had been made to widen the road to accommodate the anticipated traffic. Or looking at it another way, the trees, which have adorned the roadway for a century or more, would have to be removed so that a group of people could watch a golf match for a few days. The decision seemed to hinge on whether it would be good business to assume that golf spectators would prefer to leave for the golf tournament a few minutes earlier and plan to spend the additional driving time on a narrower road graced with live oak trees. The trees are still there, so it appears that in 1991 the United States won more than just the Ryder Cup.

Sentiments about tree rights are definitely on the rise. In the foreseeable future more than a chainsaw will be required for cutting down any big tree that just happens to be in someone's way. The native tree ordinance of Dade County, Florida, protects certain native trees regardless of property ownership. We can expect a lot of legal maneuvering and interpretations before we reach the point where it becomes illegal to cut down any tree, anywhere. But remember, only a century or so ago, anyone with a gun could shoot a deer, a goose, or any other animal. We now have hunting seasons and other controls for game animals and many other species.

One kind of activity that should change people's attitudes about big tree destruction can be seen in almost any developing town in the country. Big, healthy trees are removed from alongside the streets for some type of development. Then, in an ironic twist worthy of an O. Henry short story, a bunch of scraggly little trees are planted where the giants were. Well, maybe in a few decades . . .

Nonetheless, we still have a lot of impressive trees: the banyan tree in Fort Myers, Florida, planted by Thomas Edison and now spreading over the better part of an acre; the Angel Oak, near Charleston, South Carolina, with a trunk diameter greater than the height of a tall man; the towering redwoods. A tree in Athens, Georgia, is unusual for something other than its physical characteristics. The white oak of medium size stands on a back street, surrounded by a series of granite posts connected by chains. Inside the posts, amid the dead leaves from the tree, rises a piece of granite that resembles a tombstone. At first I thought I had found a special grave site, and I stopped to read the epitaph. Instead of a statement of death, however, the writing was a gift of life, apparently excerpted from someone's will: "for and in consideration of the great love I bear this tree and the great desire I have for its protection for all time, I convey entire possession of itself and all land within eight feet of the tree on all sides."

Someone had willed a plot of land to a tree. Upon pursuing the issue, I found the will had been that of W. H. Jackson, a professor at the University of Georgia in the 1880s. Jackson owned the land, had enjoyed the tree much of his life, and decided to leave the tree to no one but itself. The original tree was blown down in 1942, and in 1946 the Junior Ladies' Garden Club planted a sapling grown from one of the tree's acorns. This is the tree I saw. The Athens Convention and Visitors Bureau reports that the tree that owns itself was featured in *Ripley's Believe It or Not* as the world's most unusual property owner. Part of what is unusual, of course, is that an unemployed tree pays no property tax.

My first thoughts were that if this approach were to catch on, with all of us protecting a tree or two in our wills, we would soon have a lot of protected trees. Or if people got really ambitious, they could will entire woodlots or forests to themselves, protected from human whims. I am not sure how the property tax issue would be handled, but after a few decades we would certainly have a lot of protected habitat.

Nat Frazer of the Savannah River Ecology Laboratory (SREL) pointed out to me an even more intriguing issue. Why do we assume that for a tree to own itself someone has to state this intent in a will? One line of environmental ethics might be, instead, that all trees already own themselves and that we are being presumptuous to assume they belong to us in the first place.

Such a presumptuous attitude is of course the very foundation of many of our environmental problems today. We assume that the plants and animals of the world are ours, to be used (and sometimes abused) at our whim. Instead, we should recognize that their lives and well-being are intimately intertwined with our own. The increasing number of endangered species and the disappearing tropical rain forests attest to the problem of this mistaken attitude. Those who might take the position that human beings hold dominion over all plants and animals should consider the following analogy. Giving your teenagers dominion over the family car on a Saturday night does not give them permission to wreck it.

Perhaps we need guidelines that require justification for the destruction of any habitat, including every plant or animal that lives in it. Each development project or forestry operation would have to demonstrate that the economic gains to the community outweigh the environmental losses. People undertaking some ventures would have a hard time justifying that their financial gain was properly balanced by the loss of habitat, plants, and animals.

The tree that owns itself in Athens is one of a kind from the human perspective. But from the standpoint of the millions of species of plants and animals that live on the earth, isn't this really the way things were supposed to be in the first place?

Clearly, the issue of the rights of plants and animals, both native and introduced, is unresolved, as is the related issue of their environmental disposition. The ultimate controversy, however, comes with animals that kill people as part of their behavior.

They Eat People, Don't They?

A few years ago a journal called *Hamadryad* (a hamadryad is a king cobra), the official publication of a zoological park in Madras, India, had an article about crocodiles. The magazine's emphasis was on the conservation and ecology of reptiles in India and other parts of the world. Frequently, articles in such a publication offer environmental views in a manner and from a perspective that differ from that to which we are accustomed. We should consider the approaches and ideas of other cultures lest we become nearsighted and provincial in our view of the world. One thought-provoking series of articles in *Hamadryad* was about the saltwater crocodile.

The saltwater crocodile of coastal areas of Borneo, New Guinea, and Australia is the great white shark or Bengal tiger of the reptile world. A saltwater crocodile views a human being as one more source of body-building protein, an edible morsel good enough to deserve a bit of stalking and beguiling behavior if necessary. Sometimes these crocodiles swim in the ocean, far from land, but they often live in rivers and freshwater marshes. Their size is enormous, the record lengths being more than twenty-two feet. To get an idea of the awesome size and behavior of these creatures, consider that I saw a sixteen-footer lunge out of a murky pool and land fourteen feet out of the water, mouth opened wide enough for a yardstick to be placed between the top and bottom jaws. A large Australian man who was present at the spectacle told about sitting on the back of one even larger, so large that he could not touch the ground with his feet on either side.

People often ask what the difference is between an alligator and a crocodile. The answer is not simple; the almost two dozen species of crocodilians vary greatly in shape and size, within the restricted morph of a lizardlike animal with four legs. American alligators have broad snouts, whereas many crocodiles, such as the American crocodile native to southern Florida, have narrower ones. But the most significant and dramatic way that a few, but not all, crocodiles differ from alligators is that some crocodiles will unequivocally eat people. Fortunately, American crocodiles do not indulge in this antisocial behavior. American crocodiles, in fact, behave somewhat like alligators, which are shy and usually inoffensive. Some people consider American alligators a menace because they occasionally attack people and will unhesitatingly eat a dog. Usually, however, they will not intentionally approach a human being without some provocation, such as a female protecting her young, or some enticement, such as a handout of food. Almost all alligator attacks can be traced back to human irresponsibility. Saltwater crocodiles are definitely different from your plain mind-your-own-business reptile.

In one issue of *Hamadryad* an article titled "Bizarre Cult of the Croc" presented the following information about this intriguing creature. An English journalist noted that inhabitants of the Northern Territory in Australia "seemed almost jubilant whenever someone was taken by a crocodile." When a "mineworker met a grisly death in the jaws of a giant crocodile" in a national park, many of the residents in the Northern Territory displayed "widespread and bizarre delight" that the crocodile had been victorious. According to one report, an American tourist had actually watched from a faraway bank as the crocodile stalked and captured its victim. He said he had decided to visit the national park after seeing the movie *Crocodile Dundee*.

Why the lack of compassion by the residents? The article says they view themselves as "rugged frontiersmen carving out an existence in an untamed wilderness." They reject even the hint of a suggestion that the crocodile population be reduced in any way. In fact, many of the residents make a living selling crocodile T-shirts, paintings, hats, and carvings. One resident is quoted as saying that "business is certainly not going to be hurt by an occasional crocodile attack." The article closed by saying that the preservationist attitude about the saltwater crocodile is not based on "lofty conservation principles. Most of the residents instinctively believe that the waterways of the Northern Territory belong to the crocodile, and if men want to share them they must be prepared to take the risks." This is comparable to the image of the pioneering spirit of early America, which differs dramatically from many people's attitude today.

Too often, if an animal causes harm to people, we hold the animal responsible, without considering that we may have entered its natural environment, molested its young, or destroyed its habitat. We recall a newspaper account of a man who was bitten by an alligator while wading in a pond after hours at a golf course in Orlando, Florida. When the incident was reported, the alligator was killed.

Maybe Americans need to be more aware and tolerant of the nature of wild animals and more willing to share the environment. This suggestion does not mean we need acquiesce to the idea that people are acceptable prey for crocodiles or any other animal. But if some Australians can accept saltwater crocodiles, as the article says, perhaps we should be willing to accept all the animals of North America, few if any of which are out to make a meal of us.

Clearly, the issue of what constitutes tolerance is not an easy one, but it is one that we must continue to ponder, though easy consensus will seldom be reached. Human rights, animal rights, property rights, even plant or tree rights—all come into play.

2

Wanderlust, Cannibals, and Chemical Warfare: Amazing Animal Acts

A baby raccoon rustles the damp leaves, searching for anything that looks edible. A black beetle with an orange head abruptly emerges from the leaves, scurrying across the surface, sure prey for the alert mammal. A furry paw reaches out quickly, grabbing the insect. A loud pop is heard and smoke rises from the fingers that grip the beetle. The startled raccoon releases the bug and watches as its would-be meal escapes to safety beneath the forest litter.

Although most beetles are well protected by their heavy outer wings and body covering, some rely on tricks to catch unwitting predators off guard. The defense mechanism used against the raccoon is that of the common bombardier beetle. When in danger, the bombardier beetle squirts a liquid from sacs at the rear of the abdomen. A chemical reaction causes the liquid to reach boiling temperatures. The fluid immediately vaporizes into a smoky, gaseous substance. The loud pop accompanying the ejection works in concert with the smoke to surprise the enemy and allow the bombardier beetle to escape. It is worth the mild, temporary burning sensation on your fingers to pick one up and see the show.

We sometimes forget that common, well-known plants and animals can amaze us as much as the bizarre ones. We charge along, oblivious to the fascinating and seemingly infinite number of ways animals and plants have adapted to life on the planet, right in our own yards. In the world of insects, it might be hard to decide which group is most intriguing. But beetles should do for most of us.

Beetle Tricks

Beetles are the largest group of insects—almost one-third of a million have been described! To give some idea of the size and diversity of insects, one group includes the wasps and bees, another the grasshoppers and crickets, and another the moths and butterflies. Yet none of these other groups has as many species as the beetles. In fact, approximately 20 percent of the recognized species of animals in the world are beetles. Some beetle families are represented by literally thousands of species worldwide. Others, with unusual, specialized lifestyles are represented by only one, or sometimes a few, species. Some beetles are both specialized and cosmopolitan. For example, about seven thousand species of dung beetles have been described.

One North American beetle family consists of a single species that lives as a parasite on beavers. These beetles, the size of fleas, have no eyes or wings. The members of another family live in the nests of rodents and bumblebees. They feed on the eggs and young of mites that inhabit such places. Members of a family called the ant-loving beetles live in ant nests. They possess abdominal hairs from which they secrete a fluid that ants like to eat. In turn, the ants bring food for the beetles.

Blister beetles deter predators by secreting an oily substance called cantharidin that causes a blistering reaction on the skin. Cantharidin, which can be obtained by grinding up dried bodies of blister beetles, has been used for medicinal purposes as a counterirritant. Original production of the material was from a bright green blister beetle of southern Europe known as the Spanish fly. The drug, once considered an aphrodisiac, can be toxic if taken internally.

Some beetles are aquatic, and each species has special adaptations to living in such an environment. For example, before submerging, the predaceous diving beetle collects an air bubble beneath its outer wings. The bubble is placed adjacent to an air tube that allows the beetle to breathe while underwater. These common farm pond beetles eat not only aquatic insect larvae but also small fish, tadpoles, and salamanders. Whirligig beetles are oval-shaped aquatic predators that look like watermelon seeds swimming in a circle on the surface of the water. Whirligigs never run into each other or anything else. Their eyes are divided; an upper half sees above the water's surface, a lower half sees below. Try to catch a whirligig and you will find out just how well the beetle sees.

Finding out about what we consider the simplest of creatures can sometimes give us a useful appreciation of today's environments, because no species on earth is truly simple. But their complexity, whether physiological, morphological, or behavioral, may go unrecognized until someone—a scientist or layperson—studies and describes the organism.

Nothing Is Elementary

Basically, an ecologist is someone who specializes in the study of relationships between organisms and their environments, their surroundings, their homes. To ecologists, a fascinating feature of their studies is that the questions related to organisms and their environments (which include other organisms) are endless. Even seemingly simple and elementary questions can be filled with intrigue. This aspect of ecological research vexes some people, particularly engineers and developers who want clear, consistent explanations (often with quick-fix solutions). That is not the way the living world works.

More often than not in biology, seemingly simple ecological questions are actually complex and often have more than one answer. For example, why do some animals leave their homes, when they would appear to be safe and well fed where they grew up? Simple though the question seems, ecologists are still seeking the answer for different types of animals. (The answer is also sought by innumerable parents, but we will leave that question to sociologists.)

Sex Can Make a Difference for Ground Squirrels

Belding's ground squirrels, burrowing rodents that are close relatives of prairie dogs, have been the focus of detailed ecological studies by Kay Holekamp of Michigan State University and Paul Sherman of Cornell University. Their research helps explain a behavior frequently observed in other mammals—young males leaving their home territory. The ground squirrels live in the Sierra Nevada range of California. At the age of six to eight weeks, males leave the home burrow area, never to return. Young females, however, remain at the home site. The researchers' initial question, based on field observations, seems elementary: Why do the two sexes behave so differently?

An important part of any ecological inquiry is to know as much as possible about the habits and natural history of the study organism. Over an eleven-year period the researchers handled more than five thousand Belding's ground squirrels, definitely a large enough sample to reach certain conclusions. They trapped animals and marked them permanently with ear tags so that individuals could be distinguished. Shortly after capture, each squirrel was released at the site of capture and the investigators watched them through binoculars for many hours each year.

Belding's ground squirrels are active aboveground in the spring and summer, less than half the year. They spend the rest of the time hibernating in underground burrows. Each year females have an average of half a dozen young, which the mothers raise underground without help from the males. The researchers discovered other important differences between the sexes. Females begin breeding after their first year and live an average of four years. Males, however, do not mate until they are two or more years old, and the average male lives only two years. Thus many males never survive to reach maturity and reproduce, whereas most females do.

The investigators found hormonal differences between young males and females that led to greater activity and more adventurous behavior in the males. This finding provides a mechanistic, or what some scientists would call proximal, answer for why males leave home and females do not. But why would the two sexes be hormonally different as juveniles? What is the origin of the observed differences?

An evolutionary, or so-called ultimate, explanation for the hormonal difference, which affects dispersal patterns in the young of this species, is the avoidance of inbreeding (mating with a close relative) within populations. In support of the inbreeding hypothesis, the researchers observed that when a male Belding's ground squirrel managed to mate with many females in an area, he departed the next year to take up residence in another area. A male that was unable to mate effectively stayed in the same area the following year.

A pattern becomes apparent. Female ground squirrels remain in the general vicinity of where they are born. But juvenile males and males who have mated do not remain in the area. Taken together, these behavior patterns decrease the chance of close inbreeding.

Plausible answers can be given to the question of why male Belding's ground squirrels leave their birthplaces. But understanding how a basic biological principle, such as avoidance of inbreeding, governs behavior in one species does not mean we can necessarily apply the details of that understanding to other species. Complexity is characteristic of environmental questions, and continual ecological research is needed to unravel the endless biological mysteries that abound in our natural world. Scientists make theories, but the behavior of plants and animals often causes scientists to revise or even abandon theories. Nonetheless, principles do emerge in the field of ecology that retain some consistency across unrelated groups. Thus, animals characteristically modify their behavior when they belong to a family unit, although the mechanisms of expression may be as different as cannibalism in salamanders and the social structure of whales.

Cannibal Salamanders and Whales Have Family Values

First, the salamanders. Tiger salamanders are native to more than thirty states, ranging along the Atlantic Coast from Delaware to Florida, throughout the Midwest and Great Plains, and from Washington State to New Mexico. Tiger salamanders in the East and in California have a typical amphibian life cycle, with the adults living on land and migrating to wetlands to breed. The young, called larvae, have gills and live in the water, eating small aquatic animals. The juveniles eventually metamorphose and leave the water to take up permanent residence on land.

One fascinating aspect of ecology is the variability that exists within species. In populations of tiger salamanders living in the arid Southwest and parts of Mexico, many adults remain aquatic throughout their lives. Furthermore, the larvae in some populations become cannibalistic, feeding on other tiger salamanders. The cannibal salamanders are larger than noncannibalistic ones and have specialized structures in the mouth that aid in eating other salamanders. Cannibalism occurs most frequently when larvae develop under crowded conditions.

As if such a lifestyle were not intriguing enough, David W. Pfennig of the University of North Carolina and James P. Collins of Arizona State University have unveiled another facet of tiger salamander biology. Salamanders reared in genetically unrelated groups are more likely to develop into cannibals than are groups of siblings. That is, tiger salamanders that are brothers and sisters are less likely to eat each other. How do they know who their siblings are?

Salamander experiments conducted by Pfennig and Collins involved placing larvae in various groups. Some groups were all siblings and some were unrelated. All larvae, whether related or not, were of the same size, so the variation in body size could not be used as a cue to whether larvae had developed from the same set of eggs. Instead, the investigators hypothesized that larval salamanders release chemical cues that can be used to distinguish close kin from others. Presumably, genetically similar salamanders have a similar "smell."

An explanation of the ability of larval salamanders to detect—and not eat—their siblings resides in the theory of natural selection and the process of evolution. In simplest terms, evolution means change, and a species exhibits genetic change when the genetic composition in the population varies from one generation to the next. This phenomenon occurs constantly in virtually all sexually reproducing populations whenever offspring are produced. In other words, any generation will have a different array of genetic expressions than the previous generation. Natural selection is considered to be a primary mechanism involved in reorganizing the genes and their expression within populations of a species.

Natural selection, in concert with successful reproduction of the survivors, operates to perpetuate genetic compositions that produce desirable qualities (i.e., those most suitable for survival of the individual under the existing environmental conditions) and eliminate those that produce undesirable ones. Thus, the genes with the best chances of surviving from one generation to the next are those that are most suitable (i.e., the fittest) for the environmental situation.

The drive of an individual organism to survive and to pass its own genes on to the next generation is intense, whether or not its genes produce the most "desirable qualities." One proposed means of increasing the representation of one's genes in the next generation is a subset of natural selection known as kin selection. Briefly, kin selection predicts that organisms can increase their own genetic success by helping their relatives, or at least by favoring relatives over other members of one's species. Obviously, eating a brother or sister, who shares many of the same genes, would not help achieve this goal.

A whale phenomenon, although of a different nature from salamander cannibalism, is also explained by kin selection. Bill Amos of Cambridge University and colleagues Christian Schlotterer and Diethard Tautz of the University of München in Germany studied the biology of long-finned pilot whales in the Faeroe Islands, located between Iceland and the British Isles. Some male pilot whales exhibit a behavior that is different from what might be considered normal.

Pilot whales form large social groups called pods. The investigators used molecular techniques involving DNA analyses that reveal genetic relationships among individuals to establish that pod members were closely related, forming an extended family. A pod normally has more adult females than males. In most mammal species in which females live in groups, males disperse from the homesite before they become breeding adults, as demonstrated for Belding's ground squirrels. Such dispersal greatly reduces the possibility of genetic inbreeding within a species.

Among pilot whales, however, many of the males do not leave the family pod, which may number more than a hundred. Yet genetic studies revealed that males in a pod rarely or never breed with the females, who might be their mothers or sisters. Mating is presumably carried out when different pods encounter each other in the ocean. No one knows exactly how the nonbreeding males "help" their relatives in a pod, or exactly how the pod helps the bachelor males. Defense from marine predators or assistance in communal feeding efforts has been suggested. Ironically, the cohesive family structure of long-finned pilot whales makes them prey to the most dangerous predators of all, humans. Approximately seventeen hundred of these whales are killed each year because boats can be used to herd pods into coastal areas where they can be killed.

Signals

Recognition of one's species or even close kin is of key importance to most if not all animal species. An intriguing, often baffling, feature of some species is their reliance on senses we do not possess, which may be used in species for kin recognition or simply to carry out routine behavior patterns. Scientists went for centuries without understanding how bats know where they are going at night. Bats vocalize with ultrasonic cries and then detect the location of objects, including insects, by listening for the echoes. This phenomenon, called echolocation, is the same one used by whales and dolphins, which emit high-pitched signals that operate on the same principle as sonar. That is, the signal is emitted from the animal and bounces off an object. The time (in milliseconds, or thousandths of a second) that it takes to return from the object to the sender is dependent on the distance. Thus, a flying bat can determine that one moth is 20 feet away and another is only 10 because the signal takes twice as long to return from the first as from the second.

One problem with understanding the sensory mechanisms of some species is that they are totally alien to human beings. We cannot produce or sense the invisible chemicals called pheromones, which are released by insects to communicate with others of their species; therefore, discovering these chemical modifiers was not intuitive. In other cases we may be unable to produce the stimulus but be able to detect it. For example, a sensory channel used by some aquatic animals involves electric fields. The contraction of muscle tissue in animals generates a slight electrical charge, and in some species specialized cells derived from muscle tissue are arranged to form electric organs that can generate much stronger electric currents.

The pinnacle of this potential is found in the torpedo ray, which can generate about 500 volts, accompanied with an amperage as high as 2, resulting in a power output (in watts) of 1,000. Torpedo rays use their power in the sea to great advantage at feeding time, as the shock can be lethal to small fishes. According to one report, the electric shock causes the muscles of a small fish to contract so violently that its spine may snap. South American electric eels, which can reach a length of seven feet, can deliver a shock of 600 volts but at a much lower amperage. Electric catfishes of Africa also generate powerful electrical currents. Some sharks and rays that do not generate electricity do possess receptors that enable them to detect weak electrical fields. They are able to find flatfish buried in the sand by sensing the weak but steady flow of electricity from muscular activity of the prey. Detection of electrical fields is critical to some fishes that live in dark or turbid waters. By sensing changes in the electrical field, the fish use electrolocation to detect objects they cannot see.

The Shocking Behavior of Knifefishes

A study by Peter K. McGregor of the University of Nottingham and G. W. Max Westby of the University of Sheffield has added a layer of complexity to the use of electricity by fish. Although people have long been aware of the power generated by some electric fishes, the findings of the two British investigators provide evidence of a previously unknown ability of this unusual group of organisms. Scientists already knew that electric fish can differentiate their own species from others and even determine the sex of another individual by the electrical current. But McGregor and Westby demonstrated that South American knifefishes, which produce a weak electric current, use their electric receptors to discriminate between individuals of their own species.

A knifefish is a solitary creature in its native habitat, living six to ten feet away from its nearest neighbor. Unlike electric eels that can stun or kill prey, a knifefish's electric current is too weak to be used as a weapon. But a knifefish will vigorously defend its territory from others of its species by biting intruders.

The scientists employed a technique used by ornithologists to study birds that display intense territoriality against members of their own species. A bird's agitation when an intruder is in the vicinity is reflected in the intensity of its song.

Although the songs of different members of a bird species may sound alike to us, experiments have demonstrated that almost all territorial birds can distinguish one individual from another. The experiments with birds are conducted by first recording the songs of different individuals. Playing the recording of a bird from a neighboring territory elicits a weak response call. If the song of an unfamiliar bird of the same species is played, however, the response call is much stronger.

Whereas birds use sound as a means of communication, knifefishes use electric currents. The researchers placed electric fish in a large aquarium, using plastic mesh screens to set up equal, adjacent compartments. Over a three-month period, each knifefish was allowed to set up a territory and assume dominance of its compartment. Each defended the border of its territory. Some, even after three months, did not gracefully accept the presence of the neighboring fish but eventually continued to interact with electrical signals rather than with physical violence. The investigators then recorded the electric pulse of each individual so that electric charges could be "played back" using an electrode placed in the aquarium. When the scientists exposed fish to electrical recordings of familiar neighbors, they showed little response. But when the current of a stranger was used, sure enough, the knifefishes displayed strong territorial behavior. They were able to distinguish one individual from another by the fine details of the electric waveform it generated, which varied in a subtle but apparently perceptible way from that of other individuals.

Bat's Night Out

The complexity of communication and sensory abilities in the natural world is remarkable. Animals do not react to stimuli and to their surroundings like biological robots, but their behavioral responses are often keyed to subtle differences in stimuli that are not readily detectable to the human observer. And, whereas intraspecific communication is critical in some situations, interspecific awareness is vital in others. Through use of its echolocation, a bat flitting into view beneath a street lamp at dusk can snare a beetle in midflight. A minute later the same bat, a member of the world's only group of flying mammals, may dart back to the scene toward the same swarm of insects hovering around the light but this time with a different target: a flying cricket. Still hidden by the outer darkness, the bat closes in for the kill. Suddenly the cricket sticks a hind leg in front of its own wing on the side away from the approaching bat. The maneuver breaks the downward stroke of the wing. The cricket turns sharply to that side and goes into a power dive toward the ground. The bat misses. The cricket escapes.

How did the cricket know to get out of the way? Was it an accident? Absolutely not. The night is filled with sounds we cannot hear. Those sounds, ultrasonic signals from bats, occur in a range far beyond human hearing capabilities. But many night-flying insects can hear the ultrasonic signals emitted by these nocturnal navigators of insect destruction. And when they do, they take evasive action.

Bats have eyes that can see in the light, but in an unlit cave or the darkness of night they use their ears. Their navigational abilities by means of sound, not sight, were discovered almost two hundred years ago. As with many scientific discoveries, the observations were not accepted because they were not understood. By using echolocation, bats not only navigate in the dark by determining distances but also identify objects by their shape. Bats' ability to consume vast quantities of insects on the wing is well known. But all animals evolve over time, and it has become apparent that many species of flying insects have developed early-warning systems to detect the approach of a hungry bat. On the basis of his research at Cornell University, Mike May has shown that the relationship between bats and some insects is like air-to-air combat between fighter pilots who have highly sophisticated aerodynamic engineering equipment.

In the 1920s scientists observed that some moths can detect the presence of bats by sound and institute evasive maneuvers. The phenomenon was tested experimentally in the 1960s. Through the use of photography in the laboratory, scientists observed moths moving away from an artificially produced ultrasonic sound when they were several feet from the source. If the moths were closer, they went through a variety of acrobatic antics, a tactic that might be effective in eluding an attacking bat.

Insects that can hear have the equivalent of ears, in the form of flat tympana, on their bodies or legs. Hearing serves many functions for insects, as it does for other animals. Male crickets, katydids, and grasshoppers make chirps or other sounds to attract females for reproduction. Honeybees exchange information through variations in their rate and level of buzzing. But these sounds are for communication within the particular species. Being able to hear the approach of a predator that emits a high-pitched sound inaudible to human ears is a special trait presumably reserved for insects that fly at night.

Some praying mantises may hear for only one reason—to detect when bats are on the prowl. In fact, praying mantises were not even known to have hearing organs, because they show little awareness of sound in the range that humans hear. Thus, scientists assumed they could not detect sound waves.

But they can hear an approaching bat. When a mantis in flight detects a bat closing in for a meal, the insect shifts the position of its legs, head, and abdomen. Within a millisecond the mantis turns abruptly or goes into a spiraling dive. A simple moving target becomes a whirling, unpredictable one. More times than not, the bat misses.

In a contest between a bat and an insect, one is the winner and one the loser, and the cost of losing for the insect far outweighs the cost of losing for the bat. Among some insects, however, confrontation can be a life-or-death proposition for either of the combatants. For example, who would win in a contest between Japanese giant hornets and honeybees? The hornet can crush the honeybee with its mandibles. The honeybee can kill the hornet through overheating, by surrounding it en masse. The winner is determined by the homeland of the bees.

The Unsocial Behavior of Social Insects

The giant hornet is the only species of hornet known in which individuals assemble and then attack other social bees or wasps en masse. A study reported in 1995 by researchers at Tamagawa University in Japan revealed the fascinating behavior displayed by this extraordinary predator and the responses of its prey species, one of which doesn't take too well to being bullied.

The hornet's foraging strategy sounds like a four-phase military exercise. The first step in the giant hornet's feeding sequence is hunting. Hornets fly solo missions in search of nests of other social bees and wasps. Upon finding a bee's nest, the lone hornet crushes individual bees in its jaws. Dead bees are taken to the hornet nest to feed larvae. A solitary raider may return to the bee colony several times to take additional bees. Phase two involves recruitment. The hornet rubs secretions from a special gland onto the area surrounding the honeybee nest. The secretion is a pheromone, a chemical compound used in communicating specific messages to its own or other species of animals, often provoking a behavioral response of some sort. Being able to detect a pheromone is like having a sixth sense for chemical awareness—and the chemicals deliver information. In this case the hornet's pheromone is a signal for other giant hornets to amass and attack. Nestmates of the giant hornet flying in the area congregate as they sense the pheromone. Then they attack; the well-named slaughter phase is under way. As many as forty European honeybees are killed per minute, and an attacking force of twenty to thirty hornets can kill thirty thousand bees in three hours. After the hornets achieve such an impressive victory, they enter phase four: occupation of the bee nest. For more than a week the hornets carry bee larvae and pupae to their own nest as food for the hornet larvae.

The attack-and-conquer approach in Japan works in favor of the marauding giant hornets when the prey is a colony of European honeybees. This is the same species found in North America, where they were also introduced from Europe. European honeybees appear virtually defenseless against hornet mass attacks, in part because they are oblivious to the impending onslaught. That is, they are unable to detect the hornet pheromone and so are unaware of the presence or plan of the hornets.

But if you are one of those people who like to see the underdog win occasionally, stay tuned, for not all bees in Japan are the introduced European honeybee. Native Japanese honeybees can detect the hornet pheromone, and they understand the message being sent, a trait that is akin to deciphering an enemy code. When the pheromone is detected, the bees modify their behavior by increasing the number of defenders at the nest. The first hornet to attack is greeted by a swarm of more than five hundred bees, which form a large ball around the intruder. The ball of bees may stay intact for up to twenty minutes, with the inner temperature of the ball reaching 116°F, which is lethal to the hornet. The bees can withstand temperatures up to 122°F, although some bees in the center die from hornet bites, not stings.

Giant hornet attacks occur in autumn when a surplus of food is needed to feed developing hornet larvae. Sometimes the Japanese honeybees do not defeat the hornet if the surprise attack is carried out quickly, before the bees can mobilize. If the bees have a small colony, they are unable to amass enough defenders to kill all of the hornets. Even then, most adult bees usually escape and forfeit only the nest.

Giant hornet attacks and the differential response of nonnative and native honeybees are a fascinating example of coevolution, in which two species evolve in response to each other. The hornets have evolved a mechanism for acquiring large quantities of baby food in a short time. Meanwhile, the native Japanese honeybees evolved a counterstrategy, developing an effective defense against the predator. The introduced European honeybees did not evolve in a system requiring such a response, so as a species they are essentially helpless against the warlike tactics of the giant hornet.

The special world of pheromones gives insects communication powers that we would never comprehend except through modern chemistry. But another trait among insects has long been known: the power of the ant. For its size, an ant is something like a thousand times stronger than a human being.

A discovery in 1992 involving the Saharan silver ant should add to the ant's reputation as a remarkable creature and impress almost anyone interested in the extremes that ants are capable of withstanding.

Silver Ants Sup When the Temperature's Up

When the sun rises above the Sahara, most animals head for shelter, usually underground. Each day, however, a few insects and other invertebrates die because they fail to escape the rays of the rising sun. This is not an unusual phenomenon in nature; animals die all the time, and other animals eat them. Enter the silver ant. Like many other ants, these are scavengers that feed on the carcasses of invertebrates. But a big difference between silver ants and other ants is their sense of timing. Silver ants forage during a hot part of the day. In fact, whereas other ants disappear from the surface before temperatures in the summer reach 113°F, silver ants do not even leave their burrows until surface temperatures are above 115°F. Such extreme temperatures would be lethal for most animals.

One ecological question is, why do silver ants feed only at such blisteringly high temperatures? Why not *start* foraging when temperatures are a little cooler? An earlier start would allow more time to find food. Ridiger Wehner and Sibylle Wehner of the University of Zurich and A. C. Marsh of the University of Namibia investigated factors that restrict the foraging activities of the silver ants.

The highest temperature that individuals can withstand and still recover from behaviorally and physiologically is called the critical thermal maximum of a species. Silver ants hold the record, enduring temperatures on the floor of the Sahara that are above those recorded for any other terrestrial animal. Few land dwellers, which can ordinarily escape thermal extremes by retreating beneath the surface, can survive when body temperatures rise much above 100°F. But silver ants discontinue foraging when their body temperature approaches 128°F!

Because silver ants get a late start in the morning, waiting until surface temperatures are already too hot for most animals, their foraging time is limited to only a short period each day. By observing the daily life of ants and other organisms on the desert floor, the investigators discovered the answer to why silver ants restrict their foraging time. Silver ants begin their search for food at a time of day when another animal stops. By being on the surface only at midday, silver ants avoid predation by a particular species of desert lizard. Although the lizard can withstand fairly high temperatures itself, it retreats from the surface when the surface temperature reaches about 115°F—the temperature at which silver ants become active. The behavior of the ant-eating lizard has apparently shaped the ant's behavior by defining the lower limit of a thermal window within which the ant can safely forage. The upper thermal limits of silver ant activity are set by potential heat stress. The sands heat fast in the Sahara, and even a silver ant can get too hot.

How do ants in the colony know when it is hot enough for the lizards to be gone and, therefore, safe for them to leave the burrow? The researchers observed that each day a few ants were active on the surface around the entrance of the colony, apparently monitoring the temperature. When a suitable surface temperature was reached, those on the surface signaled those in the burrow, triggering the emergence of the silver ants. As is commonly known, social insects such as ants and bees have remarkable abilities to communicate. Silver ants may use secretions from glands in the head to deliver the message.

Silver ants stake their lives on attaining the critical temperature that signals the disappearance of the lizards. Once the message to emerge is received, all the ants surface at once and swarm over the desert floor in the vicinity of their nest in a speedy effort to find food. Their foraging time is short; they return to the burrow within minutes to avoid the fast-rising desert temperatures. Although the threat of the lizard predator diminishes with a rise in temperature, the threat of heat stress in the Sahara increases.

Boris Karloff of the Desert

Thousands of miles from the Sahara, another denizen of the desert emerges from its underground burrow. Once called the Boris Karloff of the desert and described as "ugly as sin itself," the Gila monster is the largest—and only venomous—lizard in the United States. The general behavior of Gila monsters is poorly understood. A study in which several of the lizards were equipped with radio transmitters revealed that they spend 95 percent of their time underground. Do they come aboveground only to search for food or find a mate? And what do they do all that time beneath the earth's surface? Some scientists speculate that a Gila monster may be able to go a year or more with no food or water, but no one knows how long for sure. Another mystery: Where do Gila monsters lay their eggs? Although this lizard is certainly conspicuous when aboveground, a Gila monster nest has never been found in the wild.

Gila monsters are fat creatures, as lizards go, with skin that looks like a covering of orange-and-black beads or pebbles. According to the book *Gila Monster: Facts and Folklore of America's Aztec Lizard* by David E. Brown and Neil B. Carmony, with drawings by Randy Babb, Gila monsters have "Halloween-hued skin the texture of Indian corn." The large, rounded tail is half the length of the body and serves as a storage compartment for fat and water. In contrast to most lizards, the tongue is forked like a snake's. The head and body are covered with a primitive armor of bony plates beneath the skin, making it almost impenetrable by the teeth of predators.

The geographic range of Gila monsters is primarily in the Sonoran desert of Arizona and farther south into the state of Sonora to northern Sinaloa in Mexico. South of the U.S.-Mexican border their range overlaps with the world's only other venomous lizard, the Mexican beaded lizard.

The Gila monster suffered due to human ignorance from the time of its scientific discovery in 1869 until the mid-1900s. During the past few decades attitudes have changed, thanks to educational efforts of environmentalists and ecologists. This fascinating animal is now recognized as part of the natural heritage of the Sonoran desert and symbolic of native wildlife of the Southwest. In the early 1950s Arizona passed a law protecting the species, the first state law to protect a venomous reptile. The species is now officially protected throughout its geographic range. And in 1985 the Gila monster missed by only a few votes (to the Arizona ridge-nosed rattlesnake) of becoming the official state reptile of Arizona.

Many myths and much foolishness center around the Gila monster. Even biologists may have a few misconceptions. For example, Gila monsters are recognized scientifically as one of the two species of venomous lizards in the world, yet no confirmed record exists of a healthy, and sober, person dying from the bite. Old newspaper articles give sensational, overstated accounts of deaths from Gila monsters. When one delves into the medical records and other evidence surrounding reported lethal bites, however, the cause of death is often attributed to the victim's major intake of alcohol, a poison different from lizard venom. In one report, a Gila monster had been falsely accused for a death presumably caused by a rattlesnake.

Unless complicating factors are involved, a person is not going to die from the venom of a Gila monster. But a bite would hurt and might scare a person almost to death. Gila monsters are noted for their tendency to "hang on like a bulldog," so the first order of business is to get the animal off. Using pliers or a screwdriver is the technique most often mentioned for prying off an attached Gila monster. The vicious teeth, which look like pieces of broken glass, can slice and tear skin if the animal is yanked off.

Is It Venomous or Poisonous?

Scientists continue to reveal a never-ending array of mechanisms animals use to communicate or to defend themselves from each other and their environment. Many organisms use venoms and poisons for defense. Catfish have venomous spines. Common garden toads secrete distasteful toxins from skin glands. Rattlesnakes inject venom through fangs, and skunks can deliver a potent smell.

What is the difference between *venom* and *poison?* A venomous animal produces a chemical substance in special glands and then *injects* it into another animal. Injection can be by any of several means, including the spines of a fish, the tail stinger of a scorpion or hornet, or the stinging hairs of some caterpillars. In any case, the venomous animal forcibly puts a toxic chemical inside the body of another animal. A poisonous plant or animal, on the other hand, produces a chemical substance that is destructive only when an animal comes in contact with it, such as by touching, smelling, or eating the poison. For example, poison ivy produces an oily substance that can produce blisters on some people upon contact with the leaves, stems, or roots. Death angel mushrooms and poison hemlock produce chemicals that are harmful to people who consume them.

Throughout most of the twentieth century, scientists could say that no birds were known to use chemical warfare for defense. In fact, one of the attributes (besides having feathers) that set birds apart from the other groups of vertebrates (fish, amphibians, reptiles, and mammals) was that they did not use poisonous chemicals as a means of protection from predators. At least that is what we used to think. No venomous birds have yet been discovered. But to the surprise of scientists, John P. Dumbacher of the University of Chicago along with several associates discovered that three species of New Guinea birds called pitohuis have poisonous skin and feathers.

The chemical composition of the poison is similar to that found in the skin of dart poison frogs of Colombia, South America. Dart poison frogs secrete a toxic chemical that makes them inedible. The toxic material is a type of alkaloid that, if eaten or injected, has an immediate effect on the nervous system. Colombian natives put the poison on the tips of darts used to hunt prey. Until the report about the New Guinea pitohuis, this particular toxin was unknown anywhere else in the animal kingdom. The chemical is a powerful deterrent, and poison frogs are avoided as a source of prey by predators. Presumably the poison operates in a similar fashion for the New Guinea pitohuis by discouraging such typical bird predators as snakes, other birds, and mammals.

The most toxic of the New Guinea birds examined is the hooded pitohui, a small orange-and-black bird with a crest like a tufted titmouse. The birds produce a foul smell, which as far as scientists know is unusual among birds with the exception of the putrid smell of vultures that have dined on rotten flesh. While collecting and preparing the specimens of hooded pitohuis, the investigators suffered from bouts of sneezing along with numbness and burning of the mouth and nasal lining. As is often true with scientific discoveries, the local populace already knew about the phenomenon. A 1977 book on folklore of the Central Highlands Province in Papua, New Guinea, mentions that local residents said the skin of the hooded pitohui "is bitter and puckers the mouth." They referred to it as a "rubbish bird" and advised that it not be eaten "unless it was skinned and specially prepared."

The findings about poisonous birds are significant in several ways. First, any increase in our knowledge of the natural world is of value in raising our intellectual consciousness. Also important is the confirmation that unrelated and geographically separated animals, such as Colombian frogs and New Guinea birds, can independently evolve the same chemical defense. Such discoveries help us better understand the variabilities and similarities among organisms.

Scientists have been studying natural history for decades, yet we are a long way from knowing it all. New traits, new behaviors, new species are still being discovered. And each species we examine in depth has traits and properties different from all the others.

3

Why Did the Turtle Cross the Road? Research Questions and Occasional Answers

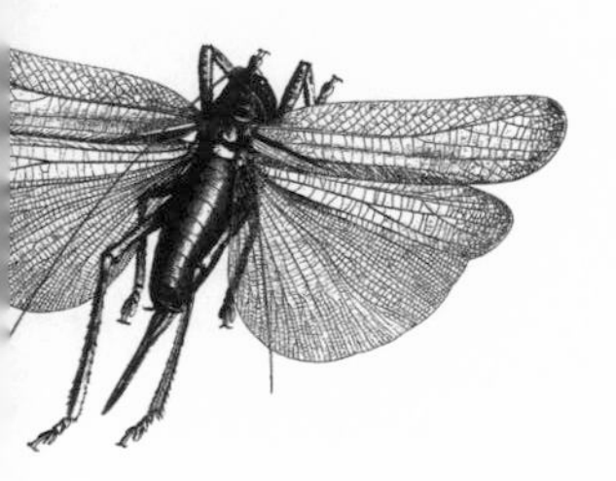

In spite of video games and entertainment TV, which keep us indoors, away from nature, Americans are better educated about environmental issues and the science of ecology than were earlier generations. We can even hope that the next generation will manage the Earth's natural habitats and wildlife in a proper manner. But today's young people, especially future ecologists, as well as many adults need to be aware of an often unrecognized problem: scientists know far less about the basic ecology of our native plants and animals than the public thinks they do. Having a thorough ecological knowledge of the life around us will be vital for those who inherit the responsibility for the custody of the natural ecosystems we enjoy. Mysteries in ecology are waiting to be solved. Lest we become smug and complacent about how environmentally sophisticated we are, consider our current level of ignorance about some everyday animals, even an animal as familiar as a turtle.

Most people have seen a turtle crossing a road at one time or another. The box turtle, the high-domed kind that can completely close its shell, is still common in some regions and is the species seen most frequently on land. Box turtles are terrestrial, so when we see one on a highway we can assume it is just walking to the woods across the thoroughfare. But what about aquatic turtles such as snappers, pond sliders, or painted turtles? All of them cross roads, too. We assume they are leaving their home in a lake, stream, or pond. But where are they going? And why? At certain times during the year, the most obvious explanation might be that a female turtle is looking for a place to lay her eggs or is returning from having done so. If it is a male, juvenile, or female without eggs, an aquatic turtle may be traveling from one body of water to another one nearby.

Scientists at the SREL in South Carolina, including Justin Congdon and Judy Greene, have been involved for years in studies that explain where aquatic turtles are headed when they leave the water. During a drought when many regional wetlands may dry completely, turtles may travel overland a mile or more to reach places where water still exists. In the 1970s the usually aquatic mud turtle was discovered to leave the water in autumn and move to selected areas on land to hibernate during winter. In the 1990s Kurt Buhlmann, a graduate student at SREL, discovered that the unusual chicken turtle, which lives in natural wetlands in the southeastern United States, also moves to land for winter hibernation. Several ecologists at SREL and other places have demonstrated that male turtles move across land from one pond to another, presumably looking for mates. Although these findings explain to some degree why a turtle might be found crossing a road, we still do not always know.

Most of what we do know about aquatic turtles moving around on land was virtually unknown to ecologists a decade after astronauts had walked on the moon. Then, in a 1992 study, Rebecca Yeomans of the University of Georgia determined experimentally that yellow-bellied slider turtles have a water-finding ability. Turtles removed several miles away from their home ponds, into areas where they have never been, can still find their way, almost unerringly on sunny days, to the nearest body of water. If released on land, they do not just wander aimlessly if the sun is shining but can walk directly through woods and across fields to the nearest water. They have a more difficult time on cloudy days. But a mystery remains: No one knows for sure how they do it. Do they smell water? Do they hear frogs calling? Do they look up at the sky and somehow perceive light reflected from the surface of the water, which would explain their greater effectiveness when the sun is out? One thing is certain, a turtle crossing a road knows where it is going. Nonetheless, after decades of biological studies, we remain ignorant about how the turtle knows where it is going and the shortest distance to get there.

Why is understanding how a turtle finds its way from one body of water to another important? To appreciate the effect of human environmental impacts on species, we must understand the capabilities of species under natural conditions. We should encourage the revelation and explanation of any natural phenomenon, whether fascinating or routine. The more we know of an animal's ecology, the more we will appreciate the species—and the better suited we will be to make educated recommendations for making human activities compatible with the conservation of wildlife.

What Else Is on the Other Side of the Road?

Another turtle mystery involves the chicken turtle. The chicken turtle has both a physical characteristic and a behavior quite different from other turtles in the same area. The neck of a chicken turtle is as long as its body—twice as long as that of other turtles the same size. Why do chicken turtles have long necks? One might think scientists would understand the function of something as obvious as an abnormally long neck in a common native species. But no one knows for sure why chicken turtles have long necks. Kurt Buhlmann of SREL and Jeff Demuth, a student from Southeast Louisiana University, determined that the diet of chicken turtles consists of a variety of aquatic insects, crawfish, and fishing spiders that hunt along the water's edge. Do the turtles' long necks help in probing into crawfish holes or in striking out at fast-moving insects and spiders?

The egg-laying behavior of chicken turtles is another ecological riddle. North American turtles typically lay their eggs in late spring and summer. Sea turtles come ashore on the beaches on summer nights. Snapping turtles, painted turtles, and pond sliders normally lay their eggs from April into July. Chicken turtles are very different. Most lay eggs on warm days in fall or even winter. This is a true ecological conundrum. Why does one species of U.S. turtle lay its eggs at the opposite time of the year from the others and never lay them during the months when other turtles nest? Such a natural phenomenon merits explanation if we are to thoroughly understand and be informed about the world around us.

Turtle, Albatross, and Spider as Senior Citizens

Among the attributes of turtles for which data are available, based on long-term field research programs and records from zoos, is the trait that individuals of several species have been documented to live longer than almost all other vertebrates. But mystery abounds, for no evidence has been uncovered that turtles show the conventional signs of aging characteristic of most species. Understanding why they do not show aging signs could be of significance to our own species. Americans are getting older every year—not just each one of us, but the population as a whole. The entire age structure of our population is shifting; people are living longer. But no matter what we do, old age eventually catches up. Even the healthiest, most active individuals eventually begin to show signs of senescence. Physicians are successful at keeping some of us alive for longer than we would have lived half a century ago, but scientists do not understand the basic process of aging. In fact, whereas some animals show senescence in a manner comparable to humans, turtles and a few other species show no signs of aging whatsoever, even at what would appear to be "old age." Scientists do not know why, and the prospects for learning about the aging process are intriguing.

The National Institutes of Health (NIH) provides funding to support human medical research, including the study of aging phenomena in other animals if the research seems likely to contribute to understanding senescence in humans. Practically all the funds are reserved for research on white rats and mice, on primates such as rhesus monkeys and chimpanzees, and on laboratory-reared fruit flies because much is known about their genetics. But in 1989, a meeting was held at NIH in Bethesda, Maryland, to consider some options for funding aging research on other species of animals as well. The meeting consisted of about twenty scientists, including ecologists, who work with animals that may provide some unique approaches to the study of aging. Although few animals have the physiological similarities to humans that rats, mice, and primates do, studying other types of animals could lead to a better understanding of the aging process—and why it occurs in some animals, like human beings, but not in others.

I provided the information about the aging, or lack of aging, characteristics of turtles. Robert Ricklefs of the University of Pennsylvania presented information about a variety of seabirds, including the albatross. Individuals of at least one of the many species of albatrosses, the royal albatross, live at least fifty years, maybe even a hundred for all anyone knows. Yet albatrosses do not show any of the typical signs of senescence so common among human beings. For example, a female royal albatross begins laying about one egg a year at the age of ten years and continues to do so year after year without appearing to get older. Although scientists do not yet know how long albatrosses live beyond fifty years, presumably if an albatross reached an age of seventy-five it would be as reproductively fit as a youngster of forty-five or fifty, showing no signs of reproductive failure. Clearly, the albatross holds some secrets we might like to know more about.

We were amazed to hear from Steve Austad of Harvard University about some of the spiders. Most spiders live a couple of years—not a surprising feat. But some of those in one group, which includes the tarantulas, have been known to live for thirty-five years. One reason tarantulas may be a useful research tool is that the males and females differ considerably in their longevity patterns. Upon reaching maturity, the female tarantula continues to molt, grow, and reproduce. The male, on the other hand, upon reaching spider manhood at age ten, mates once or twice and then dies within a few weeks or months. Humans parallel this relationship in that females live slightly longer than males, although the biological reasons are entirely different, as is the age differential. The dramatic contrast in mortality rates of the sexes in tarantulas offers an opportunity to identify a specific cause of the consistently earlier death of males. Perhaps a lesson for people will emerge.

Doddering Old Possums

Austad also presented data about another animal, the opossum, that may provide useful information in understanding senescence and the problems that accompany old age. In contrast to some spiders, seabirds, and turtles that are now known to have life spans far beyond what might be expected, opossums are short-lived. The Methuselah of the possum world was only four and a half years old. An average possum has a life span of less than two years. A doddering old possum of two and a half years in age is likely to have cataracts and show reproductive senility.

The advantage of using the possum for studying aging phenomena is that it has a lot of babies and is easily reared in captivity. Thus, within two years the researcher has a population of old individuals that can be studied for some of the same aging characteristics experienced by humans. The advantage of the longer-lived species for research is that they have somehow overcome aging problems. The job of the scientist is to discover why. The job of NIH is to recognize that white rats and fruit flies do not hold all the answers. We are not sure that NIH has reached that conclusion yet, but time will tell.

Pasteur Was Right

For some people, scientific research of any sort is most readily accepted only if part of the justification is that a direct application to human beings is forthcoming. Despite this attitude, an objective of acquiring basic knowledge of scientific principles and processes is vital to the intellectual health of any scientific field. The goal of basic research is to understand our natural world, to provide long-term contributions to human knowledge about a subject. Basic research forms the foundation for applied research, which is research directed toward solving a specific problem that is important to society.

Some people, including far too many politicians who are looking to the next election that may be only months away, see no practical value in basic research. They want research to be directly applicable to current problems and to be a quick-fix solution during their own term of office. The concept may sound reasonable from a short-term perspective, if we could always predict what our problems will be. But we cannot, so all research that has application to a current problem must be anchored in earlier basic research that was performed solely for the advancement of knowledge. Louis Pasteur stated in the mid-1800s: "No, a thousand times no; there does not exist a category of science to which one can give the name applied science. There are science and the application of science, bound together as the fruit to the tree which bears it." In our opinion, Pasteur's philosophy is still sound well over a century later.

Much of today's so-called applied environmental research might be better described as "crisis" research. One reason for a crisis attitude is that the necessary foundation of basic ecological knowledge has not been gathered. With a little foresight and proper planning (such as support for basic ecological studies), it might already have been available. For example, could we perhaps have found a timely solution to the rapid spread of Lyme disease if we already understood the natural history of the ticks that transmit the disease to humans?

If ecologists had spin doctors as effective as those in politics, many basic ecological studies could be communicated to the public in a manner that would make them justifiable as a service to humanity. But the first step after ecological knowledge has been acquired is to communicate with those in the scientific community who study the environment and who teach others about the findings. When researchers communicate their findings or justify their research plans to other scientists, the material is often esoteric and theoretically based with no reference to human problems or issues. In the field of ecology, almost any research is at least indirectly applicable to humans in that a more thorough understanding of the complexity of ecological systems results. However, making one's research relevant to society is not always the ecologist's objective.

How do research ecologists communicate with one another? Where do they get the facts on which they base their opinions, take positions, and make recommendations? How do they communicate their findings to other ecologists so that new knowledge becomes available in a timely fashion? The Internet offers rapid dissemination of some types of scientific information, but the most reliable and authoritative sources for transfer of ecological facts remain unchanged: technical journals and scientific meetings. The largest and most important annual gathering of professional ecologists, whether they are teachers, researchers, or administrators, is typically the meeting of the Ecological Society of America (ESA).

Zooplankton and Rats Make Their Reports

The theme of the seventy-ninth annual ESA meeting held at the University of Tennessee in Knoxville in 1994 was science and public policy. The issue has confronted scientists for decades. Ecologists must often balance objective reporting and interpretation of ecological facts against an advocacy role on environmental issues. In the eyes of some, a scientist cannot do both. But the two goals need not conflict, and many scientists today believe they have a responsibility to society to strive for such a balance.

More than seventeen hundred ecologists presented talks at the ESA meeting, most on basic research. One such study, "Diversity and Distribution of Zooplankton in Ponds on the Southeastern Coastal Plain," was conducted by Adrienne De-Biase and Barbara Taylor of SREL. Zooplankton are small aquatic animals that inhabit virtually all natural bodies of water, including all oceans, the Great Salt Lake, the Dead Sea, and every stream, river, and pond. The researchers studied eighty-eight temporary wetlands in South Carolina and found zooplankton biodiversity to be linked most strongly to how long a pond held water. The size of the wetland and even the level of past disturbance (such as draining for agricultural purposes) were of less importance in determining zooplankton abundance and diversity. Someone cynical about environmental research on animals that are barely visible to the naked eye might ask, Who cares about zooplankton? The answer should be, Everyone.

Zooplankton are vital to natural wetlands. Directly or indirectly they serve as food for most other animals living in and around wetlands. The zooplankton community can serve as an indicator of environmental conditions because the relative abundance of different species changes in response to subtle changes in water chemistry. Understanding zooplankton ecology and distribution patterns increases our knowledge of wetland habitats.

The value of such knowledge may ultimately be extended through other applications to wetland issues. For example, the knowledge that wetland biodiversity of zooplankton is influenced more by hydrological conditions than by habitat size or types of alterations is an ecological principle that can be tested for other animal groups. Birds, mammals, reptiles, and amphibians (many of them legally protected as endangered species) may follow the same rule.

Basic ecological knowledge about zooplankton could lead

to a better understanding of how to manage and protect wetland systems and wildlife. Acquisition of knowledge justifies basic ecological research. The knowledge sets a solid foundation for future study and puts us one step closer to understanding the complexity and intricacy of the natural world on which we depend for survival.

The eightieth annual ESA meeting, held in 1995 at the Snowbird Ski and Summer Resort in Snowbird, Utah, had as its theme the transdisciplinary nature of ecology, emphasizing that all events on the planet are connected and interactive. Thus, to understand the world's environments, ecologists must study them across all disciplines of science. Talks presented at the meeting included a mix of basic and applied ecological studies. For example, a so-called hantavirus causes a disease called hantavirus pulmonary syndrome (HPS), a respiratory ailment with a mortality rate of 50 percent. First discovered in May 1993 in the Southwest, HPS has resulted in the deaths of more than 150 people in the United States and Canada. The Ebola virus in Africa should alert people throughout the world that viral epidemics are a looming threat to human populations. Simply treating the symptoms may not be sufficient to control some viruses: it may be important to know the origin of some viruses to aid in prevention. As seen with HPS, the solution may lie in basic ecology.

Research has been conducted on small rodents at the Sevilleta National Wildlife Refuge in New Mexico for several years. James W. Brunt, Robert R. Parmenter, and colleagues from the University of New Mexico presented a paper at the 1995 ESA meeting based on their findings. They found that deer mice and pack rats occurred in greater numbers in the year of the HPS outbreak than in any years before or after. A link was thus established between rodents and the hantavirus that was emerging as a problem. Revealing a possible connection between the hantavirus and rodent abundance provides the practical reason some people seek for research. By knowing that the rodent population is the biological reservoir for the hantavirus, control measures can be taken. But the practical extension was made possible only because of funding for a long-term study on the basic ecology of rodents. The application to viral epidemics had not been considered in the original research plan.

I visited the Sevilleta study site with Bob Parmenter in 1991 when the mammal studies were being initiated. We talked about the problems of obtaining funding for basic ecological studies, especially on such creatures as small rodents. We also talked about the relationship between rodent populations in a habitat and the number of rodent-eating predators. In some areas of the Southwest, rodent numbers are held in check by the larger hawks, rattlesnakes, and bull snakes, along with coyotes, which have been known to eat thirty mice in a night. Eliminate snakes, hawks, and coyotes; increase the number of mice and rats. Relationships exist between soil nutrients, vegetation, plant-eating prey species, and predators. But the environmental details and how they influence each other have seldom been determined. For example, how does soil chemistry affect vegetation conditions that in turn influence the rates at which rodent populations increase? And what controls the population sizes of snakes, hawks, and coyotes that feed on the rodents?

The hantavirus exemplified the ESA meeting's transdisciplinary theme. Unraveling the intricacies of how to avoid a hantavirus epidemic will involve the medical profession as well as botanists, zoologists, and soil chemists. Such an effort is further proof that preparing for unexpected biological threats often demands a solid foundation of basic ecological information across all disciplines.

Causes and Effects

The 1996 ESA meeting in Providence, Rhode Island, had the theme of ecologists-biologists as problem solvers. I sampled the program book containing summaries of papers to see if any of the thirty-four hundred participants really did solve any problems most people would care about. I randomly opened the book to the L's.

One study involved development of a model based on ecological research to integrate economic and environmental goals in the timber industry. Harbin Li, Carl Trettin, and associates from the U.S. Forest Service Center for Forested Wetlands Research in Charleston, South Carolina, plan to use information from two often divergent viewpoints—harvesting timber and maintaining a suitable habitat for wildlife species in an area. By examining both the economic and the environmental cost-benefits of timber production programs, the approach could be a big problem solver for the forest products industry. The Northwest would be better off today had such a balanced plan for meshing economic and environmental goals been established before spotted owls flew into the spotlight.

Another study cited in the program book involved research on delayed impacts of Tropical Storm Agnes on blue crabs in Chesapeake Bay; the research revealed the fragile nature of the ocean ecosystem. R. N. Lipcius and R. D. Seitz of the Virginia Institute of Marine Science at Gloucester Point reported that blue crabs declined significantly shortly after Agnes passed through in 1972. The devastation of seagrass beds on which crabs depend for nursery habitat was presumably the primary cause. Also, soft-shelled clams, major prey of the crab, declined and have remained in low numbers to the present. As the seagrass recovered, juvenile crabs increased in abundance. However, to this day adult blue crabs remain in low numbers. The investigators speculate that the lack of recovery is attributable to too much fishing pressure and too few clams. The study reveals the vital links between major environmental disturbances, the resulting alteration of critical habitat, and the close interdependence of species. Clearly, if the environmental effects are still measurable after almost a quarter of a century, regulation of commercial impacts must be carefully considered if a sustainable crop of blue crabs is to persist.

G. E. Likens of the Institute of Ecosystem Studies, Millbrook, New York, and his colleagues examined long-term effects of acid rain and concluded that the consequences may be more serious than previously anticipated. Stream research has been in progress for thirty-three years at the Hubbard Brook Experimental Forest in New Hampshire. Changes in pH, the measure of acidity, have been relatively small in the stream. However, losses of calcium and magnesium from the surrounding soils have been high, decreasing their capacity to buffer the continued inputs of acid rain. The 1990 amendments to the Clean Air Act are designed to decrease atmospheric acid deposition. But the investigators anticipate slow recovery of soil and streamwater chemistry because of the observed changes in the acid-base levels in the soil. An extension of the findings is that forest ecosystems may be much more susceptible to degradation from chronic exposure to acid rain than formerly believed.

Without even getting into what A. Liebhold and K. Gottschalk of the Northeastern Forest Experiment Station, Morgantown, West Virginia, predict about forest susceptibility and future invasions by gypsy moths, one thing is clear about the ESA meetings. Ecologists are definitely identifying the world's environmental problems, a first step toward solving them. The L's satisfied my curiosity. I feel certain other letters of the alphabet are doing equally relevant environmental research.

Effects and Causes

Basic ecological findings not only can contribute to our knowledge of the universe, a worthy goal in itself, but can cornerstone a future scientific edifice that someone thinks is important for another reason. And indeed a cornerstone is all that we can now see for some findings, although the application to goals some think are more important often comes quickly. An example is seen in the publication of a study on butterflies, insects that are more than just another pretty face. They can let us know we are altering the natural environment—maybe too quickly. Michael C. Singer and Camille Parmesan of the University of Texas and Chris D. Thomas of the University of Birmingham in England studied rare checkerspot butterflies in the United States. They found evidence of short-term genetic change, or rapid evolution, in the butterflies' preference of host plants and diet in response to human alteration of the environment. But in Europe, habitat changes have occurred so fast the butterflies could not evolve, resulting in their extinction.

At a site in California, the butterflies once fed primarily on a plant species that was parasitic on trees. During logging operations, the plant was virtually eliminated from the site. Once the original food source disappeared, the butterflies began using another plant species that was present at the site. The new host plant was no more abundant than before the logging, but the habitat alteration resulted in its remaining edible for the butterflies for a longer period into the dry California summer. During the 1980s the butterflies colonized this new host and, although not their first choice originally, the butterflies eventually evolved a preference for it. Thus, when logging removed the long-standing host, the butterflies adjusted to the situation by laying eggs on the replacement plant.

Meanwhile, at a Nevada site, cattle ranchers introduced a European weed that proved more suitable for the butterflies than their traditional host plant. In 1983 most female butterflies preferred to lay eggs on the native host. By 1990 most preferred the introduced weed, and some even refused to accept the native host. Experiments showed that the change in preference was genetic. Thus, they had rendered themselves genetically dependent on the human modification of the habitat, a case of rapid evolution in response to human alteration of the habitat.

At first glance one might view these findings as support that animals can adjust to any environmental insult we can offer. But do not be misled. Records were also available for the same group of butterflies in Europe. Sometime in the past, many populations of this butterfly became dependent on using plants in ancient, traditionally managed hay meadows. Because of rapid agricultural changes of ancient hay meadows, many populations have become extinct. Thus, the human alteration of the habitats outpaced the evolution of the insects and possibly of other animals.

Likewise, lacewings, another type of flying insect, can give notice that pollution may be affecting animals in subtle, but permanent, ways. Geoffrey M. Clarke of the CSIRO Division of Entomology, Canberra, examined two phenomena in lacewings living near a Russian chemical plant. One phenomenon is called fluctuating asymmetry. Most organisms are bilaterally symmetrical, which simply means that the two sides of the body are almost identical. This symmetry occurs in most animals: fish, elephants, insects, and people. Asymmetry, however, is not uncommon. That is, one side of an animal may vary slightly from the other. An extreme example is that of adult flounders, which have both eyes on one side of the body. The two sides are not symmetrical, although the condition is perfectly normal. In contrast, the face of Dick, one of the convicted killers in Truman Capote's *In Cold Blood,* was skewed a bit. Dick's face was asymmetric but in this case considered abnormal, because faces are supposed to be bilaterally symmetrical. Evidence of abnormal asymmetry has been used with some species to indicate that organisms are undergoing environmental stress and genetic interference during development.

In the case of lacewings, the veins in the wings are normally in perfect symmetry, identical on each side. Clarke compared insects captured around the Russian chemical plant with those from uncontaminated habitats. He found no evidence of fluctuating asymmetry. One interpretation might have been that the lacewings were unaffected by the chemical pollution. But Clarke also used insect wings to test environmental quality in another way. He looked for phenodeviants, any deviations in a characteristic, such as wing vein patterns, that differ from those normally observed. That is, the two wings of a lacewing could be symmetrical, or identical on each side with no asymmetry, but at the same time be different from the normal pattern for a lacewing's wing.

The phenodeviant analyses revealed that lacewings had, in fact, been affected by proximity to the chemical plant. Thus, one type of analysis, phenodeviance, revealed environmental stress and uncovered evidence of developmental instability, whereas another test, asymmetry, did not. The lesson: the environmental effects on organisms, including human beings, can often be subtle and difficult to detect.

Finding evidence that native species have survived human modifications is not proof that we are not damaging the ecosystem and causing extinction. Remember, the only animals scientists can study are the survivors.

Anyone Need Some Old Tires?

Another finding in entomology could signal a threat to our own welfare instead of an insect's. Larval mosquitoes were the focus of a study by Todd P. Livdahl and Michelle S. Willey of Clark University in Worcester, Massachusetts. They examined the invasion prospects of an insect that carries a disease that affects human beings. A container-breeding mosquito has recently been introduced from Asia in imported automobile tires. (Why on Earth anyone in the United States is importing used tires from Asia, we cannot imagine. Don't we have enough already? Nonetheless, apparently we bring a few thousand tires in each year just in case a shortage develops in someone's backyard or vacant lot.) Along with one or more shipments of Asian tires came this unwanted visitor. Our native tree-hole mosquitoes normally lay their eggs in water collected in tree holes or stumps, but they will adapt to artificial containers such as old tires. However, the Asian variety of tree-hole mosquito is a potential vector for dengue fever, also called breakbone fever. This infectious viral disease causes severe headaches and intense joint pain.

The newly introduced Asian mosquito has now spread from its area of introduction in Houston, Texas, throughout much of the southeastern and midwestern United States. Although both the Asian and the native species can ultimately exist in the same region, competition does occur. And the Asian species is more successful than the native form in tire habitats, a good reason not to put off cleaning up any old tires that may be lying around.

Snails Need to Get the Lead Out

If you agree with cleaning up macrolitter but are skeptical about whether invisible pollutants in small doses, such as lead or acid rain, can really cause serious environmental problems, consider snails. You may not expect scientific proof from a snail, but research on these slow-paced invertebrates has disclosed some unexpected—and disturbing—findings.

When excessive lead is in the environment, snails, like human beings and other animals, accumulate the toxic metal in their bodies. But ironically, you might be better off eating escargot from a site where lead pollution has occurred for centuries rather than just decades. Michael C. Newman and Margaret E. Mulvey of the University of Georgia's SREL conducted a study on land snails in Great Britain. The investigators sampled common garden snails from field sites, not restaurants, in England and North Wales. In collaboration with Alan N. Beeby and L. Richmond from South Bank University in London and R. W. Hurst from Chempet Research Corporation in California, they conducted laboratory analyses to determine the level of exposure of each population to lead. That is, had the snails been living in a high lead environment for only a few, or for many, generations?

Some of the snail populations had been exposed to high lead levels for long periods of time, up to two thousand years at some of the sites mined since the Roman occupation. Others had been exposed to modern sources, such as leaded gas, for far shorter periods of only a few decades. To estimate the duration of exposure, the researchers used a technique that revealed the proportions of different types of lead at a site. Like many other chemical elements, lead has several nearly identical chemical forms, called isotopes, which have different physical properties that are distinguishable with laboratory analysis. Thus, the lead at each site could be characterized by an isotopic signature, a chemical fingerprint. Different isotopes often have different origins. By calculating the proportions of the different lead isotopes, the researchers determined how the metal at a site had been derived. At some sites, the lead was primarily from modern human sources such as automobiles. At other sites, most lead was from older mines and smelters. On the basis of this information, they estimated the time period over which a snail population had been exposed to high lead levels.

Using snails collected from different sites, they compared the amount of lead sequestered in the shell with that present in the soft body tissues. In snail populations with long-term exposure, the proportion of lead in the shell compared to soft tissues was higher than that from populations with shorter exposure periods. That is, snails of the same species sequester the toxic metal differently depending on the time that the site, and the snail population, has been exposed. For a person eating snails, lead in the shell is far better than lead in the body.

One implication of the study is that storing a toxic material (lead) in inert tissues (the shell) has been enhanced due to the continued exposure of the populations. One explanation for this development is that the difference in snail populations has resulted from genetic selection. Genetic selection could occur if snails accumulating the most lead in the soft parts of the body are less healthy and produce fewer surviving offspring than those that store the poisonous metal in the shell where it apparently can do little harm.

Meanwhile, a study on the European continent in 1994 revealed the importance of snails in considering another form of environmental pollution: acid rain. Jaap Graveland at the Netherlands Institute of Ecology and other scientists examined the ecological impact that acid deposition on soils has had during the past several years. In forests with poor soils, they observed that eggshells of birds called great tits and of other songbirds had become increasingly thinner and more porous. Concomitantly, desertion of clutches by the birds had become more common. They further established that snails are critical for eggshell production in great tits and many other bird species as a source of calcium in the birds' diet during egg production.

An environmental domino effect became apparent within the trophic structure of the forest ecosystem. The investigators documented that species diversity and abundance of snails had declined in regions with poor soils as a result of loss of calcium from the topsoil by acid rain. Reproductive success of birds decreased due to the lack of snail shells, and the result was a local decline in bird populations. Snails may be slow and their messages subtle, but the news they are delivering about the environmental condition of the world is flagrant and coming all too fast.

Fire Ants Are Illegal Aliens

With some animals, such as the imported fire ant, even the message that we may have a problem is not delivered in a subtle manner. Imported fire ants are a plague of sorts. Nobody much likes them: they have vicious stings, are hard to get rid of, and do not belong in North America in the first place. But inasmuch as they are now here, apparently to stay, we should be finding out all we can about their ecology, behavior, and genetics.

Some species of fire ants actually are native to the United States. Their numbers and dispersal are kept in check by natural environmental controls that operate on native species anywhere. The problem is the species of fire ant that is *imported*. This nasty little animal was brought (imported) into the United States at the port of Mobile, Alabama, shortly after World War I, presumably in a shipment of bananas, timber, or other commerce. With no natural enemies, the imported fire ant started a war of its own. It seems to be winning.

In their native Brazil, fire ants are not particularly abundant. But in the United States, opportunities abound. By the time of the Great Depression, imported fire ants had extended their geographic range away from the docks and into Mobile County. When World War II began, fire ant colonies were becoming noticeably widespread in Alabama, Mississippi, and Louisiana. By the 1950s the range expansion of the imported fire ant was reported to be about five miles per year. Plus, infestations were being generated inadvertently in more distant regions by cargo shipments. Today the march of imported fire ants continues up the Carolina coastline, across Texas, and into the mainland interior of the southern states.

Imported fire ants were reported in the sugarcane fields of Florida, near the Everglades, in 1970. R. H. Cherry and G. S. Nuessly of the University of Florida Everglades Research and Education Center subsequently gathered evidence on the aggressive nature of these insects. The investigators used special baited traps to measure relative abundance of the ant species that inhabited the Florida sugarcane fields. The study showed unequivocally that fire ants are now the dominant ant species, resulting in a lower relative abundance of native ant species. A similar study done in Louisiana sugarcane fields showed that between 1960 and 1987 the imported fire ant achieved "undisputed dominance" there as well.

Is there a silver lining to this cloud? Well, imported fire ants provide lots of opportunities these days for scientific research. And they may have shed some light on a question related to the behavior and genetics of social insects. In some colonies of social insects, many of the developing larvae are potential queens. Most are unsuccessful. A variety of factors can influence the outcome of what might be called reproductive competition among nestmates. Age, body size, and degree of ovary development have all been identified as important in one case or another. That is, a chosen queen may be older, larger, or more developed reproductively. But in no instance had a genetic association been established. Kenneth G. Ross of the University of Georgia extended our knowledge on this subject.

Some imported fire ant colonies contain numerous queens. In these colonies the nonreproductive female worker ants select some individuals to be queens and destroy the others. Ross conducted genetic studies on such colonies. His question was, How do the workers know whether to kill an ant or make it a queen? Are those chosen to be queens genetically distinct from those that are not? He found that reproductive queens and nonreproductive workers were genetically similar with one exception. A genetic condition absent in queens was present in other members of the colony, including the ones eliminated as queens. Presumably those with the particular genes have a different pheromonal output that influences the behavior of other ants in the nest.

Although the complete story of fire ant genetics is exceedingly complex, a hereditary basis for selection of queens was established. Research of this nature reveals the intricacy of the genetics and ecology of social insects. It also confirms that the natural history of most species is still an unsolved puzzle. The more we understand about how animals function, the more likely we are to find solutions to environmental problems, such as fire ant invasions, with which we are intimately involved.

The Snake That Went Out in the Cold

Ecologists have to start on the ground floor with all studies, especially when they study plants and animals with no obvious application to an apparent problem. Consider, for example, the question, Why do herpetologists never find snakes in the winter? University of Georgia graduate student Mark Mills and I discussed this question once—on a cold winter day, as we looked at a big cottonmouth coiled in the sun among some cypress roots. The answer has become rather obvious since that day: Herpetologists seldom look for snakes in the winter. The reason for the behavior of herpetologists is obvious, too—reptiles, being cold-blooded victims of low temperatures, cannot heat their bodies except by basking and consequently would presumably be underground during cold weather. Who would expect snakes to be out in the cold?

Snakes cannot generate their own internal body heat in the way that birds and mammals can, yet they need to be warm to move, eat, and go about other snake business. Snakes and most other reptiles are presumed to remain inactive at cold temperatures by snuggling into a hole until the weather warms in the spring. In most parts of the country, snakes do not eat during winter (or at least we don't think they do) but instead rely on the food energy stored in their bodies in the form of fat. Winter is a dangerous time for a reptile. If the animal freezes, ice crystals can burst body cells. A cold, sluggish snake may be unable to escape from a predator that digs it up. So if we do not find a snake on a summer day, we attribute it to chance or, for a herpetologist, to "bad luck." If we do not find a snake in winter, we conclude that it was "too cold." Explaining why an animal does something, such as why a reptile emerges on a cold day, can be harder than saying why it does not.

Until the 1990s no one had really ever asked the question of why cottonmouths and water snakes can be seen on cold, sunny winter days in South Carolina. In fact, each time we saw a snake in the winter, we considered it to be an anomaly. In response to the realization that we had recorded a lot of wintertime anomalies, Mark Mills included wintertime observations as part of his research project designed to look at the ecology and behavior of brown water snakes.

Soon after, John Lee began his graduate research with the objective of not only documenting how frequently cottonmouths are active in winter but also finding out what temperatures they select when given a choice. At SREL, Mark and John had the perfect place to work, the Savannah River Site (SRS). The protected nature of the large tract of land and the diversity of natural habitats make the site ideal for ecological research.

To find out the frequency at which snakes actually bask on winter days (remembering that not too long ago we were not aware of how common the phenomenon was), Mark and John observed snakes in the Savannah River and adjoining swamp.

Mark captured more than twelve hundred brown water snakes on the river during his study. Sure enough, he found that brown water snakes are often out basking during winter. Although the numbers of cottonmouths were lower, John observed that they were frequently active in the swamp on days we once thought were much too cold for a reptile.

In addition to their captures and observations, the investigators placed small radio transmitters into some of the snakes. A transmitter can reveal not only the location of each animal but also its body temperature. Thus on any given day they were able to determine the exact location of each individual snake by tracking the transmitter signals. During the winter, Mark found the water snakes in places inland from the trees bordering the river (where they are common during summer), places difficult to reach on foot or by boat, places where a casual observer (even a herpetologist) would not look for them.

For his study, John set up computerized data-collecting systems in the swamp and in an outdoor enclosure to monitor the weather, the temperatures underground, and the body temperatures of the snakes. The temperature data were recorded every fifteen minutes. In this way he was able to determine the temperatures the snakes chose in relationship to those available to them. He was also able to find out how often cottonmouths are able to reach their preferred body temperatures, assuming they prefer to be warm rather than cold. From the snakes that were ultimately outfitted with transmitters or observed in the wild, Mark and John were able to gather impressive data sets. Cottonmouths and water snakes are much more active in the winter than any of us had known.

Answering a question in natural history is often akin to cutting off the Hydra's head—two more grow to take its place.

Now we wonder why cottonmouths and water snakes bask in the winter sun, warming their bodies and using precious stored fat energy faster than they would down in the cold ground. Being warm can be costly for a reptile that depends on stored fat during winter to keep from starving to death before spring. But if such behavior did more harm than good, these winter-active snakes would not survive long enough to mate and pass on the genes for this behavior to their young. The behavior presumably serves some useful function. What is it?

Another mystery is why some individual snakes choose to warm in the sun on certain occasions, eschewing a safer, though colder, underground retreat, whereas on seemingly (to us) similar days the same individuals remain inactive. Merely knowing the temperatures reptiles select in the wild, a feat accomplished in John's study of the cottonmouth, advances ecological knowledge. Such knowledge may bring us closer to an understanding of the relationships between animals and their environments, benefit us simply as a result of the study of organisms, and create in its wake a new set of environmental mysteries to unravel.

What Good Is a 3,000-Year-Old Tree?

Two apparently unrelated groups of organisms, turtles and bristlecone pine trees, have two traits in common: Both reach fairly old ages compared to other reptiles and trees, and both produce annual rings that can be used to tell their age. Anyone who has seen a tree stump or a log is familiar with trees' concentric rings, which in some cases may represent dozens of decades. With turtles, the rings are produced on the surface of each of the plates of the shell. To ecologists, such rings are a valuable tool for looking into the past, for only a few years with turtles but for centuries with some trees.

In both cases the utility of the rings comes from their width, which reflects the individual rate of growth. The edge of a ring represents a cessation of growth, such as during winter, whereas the space between two edges is added during the growing season. Growth rates are indicative of environmental conditions experienced by the organism. For example, during an unusually cool spring and summer, the width of a ring on a turtle shell might be less than one produced during a continuous warm spell.

Antonio Lara of the Tree-Ring Research Laboratory of the University of Arizona and Ricardo Villalba of the University of Colorado measured ring increments of alerce trees from Chile to determine long-term temperature records for the region. Alerce trees are conifers that grow on the slopes of the Andes. As with many scientific studies, understanding the details of how measurements are taken and interpretations are made requires training and experience. In simplest terms, the investigators compared tree-ring widths with regional temperature records from weather stations in Chile and Argentina since 1910. The assumption was that rings of earlier years could then be used to estimate summer temperatures of the region. With alerce trees, cool summers produce wider rings than do hot summers.

Scientists who do tree-ring analysis use live trees and the stumps of trees that have been cut down. The rings on a stump are visible at the site. For living trees, a hollow metal tube called an increment borer is used to bore a small hole into the trunk. When the borer is withdrawn, a small diameter segment (or core) of the tree is removed, without harming the tree. The annual rings visible on the core are representative of the tree's growth pattern. Lara and Villalba cored living trees but also used the stumps from trees that had been illegally logged. The oldest stump was judged to be 3,613 years old. (At the time the Parthenon was built in ancient Greece, the tree would have been more than one thousand years old!)

Documenting that some trees live for literally thousands of years is ecologically significant. The research also revealed information about climate patterns over the last three millennia. For example, the researchers found the longest period of above-average temperatures to have been from 80 B.C. to A.D. 160. The coldest temperature intervals were from A.D. 300 to 470 and 1490 to 1700. The calculated departures from average summer temperatures were normally less than 4°F but were as high as 9° in some years. The important point, however, is that regional temperatures remained above or below the average for intervals varying from decades to more than a century. Such temperature deviations would have a long-term effect on growth patterns for all vegetation in a habitat.

One conclusion reached by the investigators is that the alerce tree-ring analyses do not "provide evidence of a warming trend during the last decades of this century that could be related to anthropogenic causes." In other words, based on their interpretation of growth patterns of alerce trees over the last three thousand years, they cannot identify a significant global warming trend as a result of human activities.

Tracking temperature trends for a region in this manner is fascinating, but we must be cautious about overinterpreting the statement about global warming. It in no way justifies continued atmospheric pollution and the release of greenhouse gases. The main message of the investigators was to show the complexity of global climates. While warming occurs in some regions of our planet, other regions may remain cool.

Our great dismay upon reading this paper was the realization that in 1975 some people cut down a tree that was thirty-five centuries older than they were.

4

Not All the Answers Are Black-and-White: The World of Sight and Scent

A small parasitic wasp sits in a bush. Aware of chemical cues from the millions of molecules drifting around her, the wasp suddenly becomes alert. She has sensed a wave of terpenoid molecules in the surrounding air. Others follow and the wasp leaves her perch. Tracking the airborne trail, the female wasp, a mother-to-be, finds the partially eaten corn leaf that released the chemical. A rapid search reveals an armyworm caterpillar feeding on the plant. The wasp deposits an egg in the body of the caterpillar, which will become the food source for the developing wasp larva.

The details of this complex relationship between a parasitic wasp, its caterpillar prey, and the plant on which the caterpillar feeds was discovered by Ted C. J. Turlings, J. H. Tumlinson, and W. J. Lewis of the U.S. Department of Agriculture. They found that corn seedlings fed upon by the larvae of beet armyworm caterpillars release volatile odors in the form of chemicals known as terpenoids. These odors attract an animal that eliminates the one damaging the plant. The investigators' experiments revealed that a leaf does not release the chemical odors if it is simply damaged. The odors are produced only if oral secretions from the caterpillars come in contact with a damaged portion of the leaf. Caterpillar oral secretions placed on undamaged leaves do not create the odors. The tell-tale chemical smell is produced only when a caterpillar is feeding on the plant.

The development of the chemical signal by the plant and

the ability of the wasp to detect it are in the best interest of both wasp and plant. The wasp can learn from experience to use the volatile odors to locate the caterpillars. Like many of the thousands of species of wasps known as parasitoids, the wasps being studied must lay their eggs in a living host. This particular wasp species lays a single egg and, unlike some species of parasitoid wasps, does not paralyze the host insect. But as the wasp larva develops, it feeds on the internal organs of the caterpillar. Once parasitized, a caterpillar's rate of feeding on the plant is gradually reduced and finally ceases entirely. Ever wonder where the idea for the movie *Alien* came from? This could be the answer.

The terpenoids produced by the caterpillar-damaged leaves may also attract other natural enemies of the herbivorous caterpillar, as the terpenoids clearly signal the presence of a feeding insect. However, the investigators believe that attracting the parasitoid wasp or predators to the caterpillar is probably a secondary function of the release of terpenoids, which may serve primarily to deter some herbivorous insects from eating the leaves and to prevent infections of the damaged plant by microorganisms. Although it is easy to see how these findings might ultimately be applicable to agricultural pest problems, the most important contribution of the research is that an explanation has been given of the intricate and complex structure of the natural world we observe around us.

Color Me Camouflaged

Anyone who observes nature and the outdoor world can become an amateur ecologist, although most of us would never be able to detect or even suspect stimuli as subtle as terpenoids. But observing in a questioning manner is easy when we use some of the senses on which we rely greatly. Start with color. When you see a plant or animal ask yourself, Why is it this color instead of another?

Striking colors, of presumed significance to the existence of the plant or animal possessing them, set the stage for environmental mystery. In not-so-colorful animals, stripes or spots can lead the curious to wonder why particular patterns emerge. Although ecologists themselves do not always agree on the answers, those given are often intriguing. The reasons for some color patterns seem apparent. For example, many male birds develop bright plumage in the winter and spring—the period of courtship—whereas the females remain drab. Because birds can distinguish color, the males use their plumage to attract and impress females. The breeding colors of the males in some species are also used to threaten or intimidate other males. The ready distinction between males and females assures that courtship efforts or acts of aggression are not expended on the wrong sex.

The mammals represent a group for which a general question can be asked about color: Why are almost all mammals some shade of black, brown, or white? Birds, butterflies, and flowers come in an endless array of yellows, reds, and blues. But except for a few of the primates, mammals are stuck with varying shades of dull. One reason mammals have few displays of bright color is that most of them are colorblind. Thus, the common function of color to attract or dispel others of the same species serves no role in most mammals.

Ecologists generally consider that the various shades of brown seen in many mammals are for camouflage. Camouflage is important for both predator and prey species because, in either case, the animal does not want to be seen. For example, the spotted coats of leopards and fawns help them hide from the eyes of others. Although the reason for camouflage coloring is different for leopards and fawns, both predator and prey are clearly adapted for blending into particular habitats. White-tailed deer and mountain lions are brown, a color that blends in well in their forest habitats. Some species, such as the snowshoe hare, actually change their coat from brown in summer to white in winter in keeping with the seasonal changes in color.

One problem for ecologists comes in trying to explain black-and-white color patterns in mammals. Most species with black-and-white fur have special traits. The prime example in North America is the skunk. Evidently the highly visible black-and-white contrast is not protective camouflage but a warning signal to other species. Even a colorblind bobcat can presumably learn not to tangle with what might otherwise look like an easy meal.

Other black-and-white color patterns of mammals are less easy to explain, and ecologists often do not agree on what the correct explanation is. For example, at least fourteen different hypotheses have been proposed for why zebras have black-and-white stripes. One suggestion is that the striped pattern creates moving vertical lines that are difficult to focus on by a predator such as a lion. Thus, every member of the herd can sometimes escape before an individual has been singled out for capture by a confused cat. Another suggestion is that zebras once lived in partially shaded forested areas where the stripes served as camouflage. The stripes in this case are considered holdovers from an earlier time. Whether either hypothesis, or a dozen others, has any validity is uncertain.

The color of a plant or animal usually tells an environmental story. In fact, one of the pleasures of simply walking through a neighborhood can be to look at birds, or insects, or flowers and imagine why each one is a composite of particular colors and distinctive designs.

Albinos Come in Many Colors

The environment often dictates the color pattern of organisms through the mechanism of natural selection. Green snakes amid vines and shrubbery are protectively camouflaged from the searching eyes of predatory birds. A flounder flattened against the bottom sand in an estuary defies detection by fish-eating predators of the sea. But one color phenomenon, albinism, is not a product of the natural environment of plants and animals and is not continued in genetic lines through natural selection of the expressed trait. Albinism is the expression of a genetic condition that can be inherited, although neither parent need be an albino. An albino is incapable of producing melanin, the dark pigment that normally gives color to hair, skin, feathers, and other surface tissues in birds and mammals. Because of this abnormal condition, survival becomes a difficult struggle.

Albino plants exist, but they are even rarer than albino animals. One reason for the extreme rarity of albinism in the plant world is that such plants lack chlorophyll, the pigment that makes them green and is also key to their nourishment. Chlorophyll, when in the presence of sunlight, converts water and carbon dioxide into sugar. Sugar, a plant's food source, is essential for a plant to grow and stay healthy. Once plants have used up the food stored in the seed, most cannot live without chlorophyll. A neighbor once gave me several albino buckeye plants that lived for two to three weeks as little white trees, reaching a height of a foot and having perfectly white leaves.

Albino animals have their own special survival problems, too. A species that requires protective coloration is in trouble. A solid white tree frog sitting on a cattail plant by a lake shore would certainly be conspicuous, too conspicuous for a hungry heron not to see it. Without the normal camouflage of bright green coloring, similar to the cattail, an albino tree frog is doomed. Albinos of other species face similar problems because they lack the pigments and patterns that enable them to blend into their surroundings.

Poor vision is characteristic of albinos. Lacking pigment in the iris, an albino's eyes look pink because tiny blood vessels in the eye are visible. Eye pigments are important in filtering light, hence albinos have difficulty seeing in bright light. An albino hawk, hunting at midday, would be unable to see its prey.

Another problem that would be insurmountable for some species is an albino's vulnerability to sunlight. A desert tortoise with no pigment to shield the sun's rays would soon die. Animals that grow to adulthood as albinos are usually species that come out at night or live in dark habitats such as lake bottoms or underground burrows. Species with protective coverings, such as feathers, that prevent sunlight from contacting the skin might also manage to survive as albinos. But in species such as birds, color differences allow recognition of the sexes. Thus, the male goldfinch turns bright yellow in the early spring, in contrast to the olive drab female. Although an albino goldfinch would behave in the manner of its sex, it would not have the accompanying color pattern.

Albinos have been reported in many animal species including gorillas, goldfish, sharks, robins, rattlesnakes, tadpoles, and human beings. Judy Greene of the University of Georgia's SREL caught an albino mud turtle, the only one of its kind ever found. The specimen was captured as a hatchling. Hatchling mud turtles are normally solid black on top and the size of half a walnut shell. This little creature was normal in every way except for a white body and bright pink eyes. Without a little help from humans, a white mud turtle might have survived less than a day in the wild. Most albinos have many difficulties. In natural environments, they seldom live to adulthood. Among the reasons, of course, are the perils of being too obvious to other animals. In addition, the physiological problems associated with the absence of pigment can be overwhelming in certain environments. However, human intervention, as always, can change the natural order of events. Because albinism is hereditary, selective breeding can produce pure lines of albinos. The outstanding examples are laboratory white rats and mice, the most successful and plentiful albinos in the world.

Even albinism and related phenomena in which pigment is absent can be more complex than we might first assume, in part because not every solid white animal is an albino. Suppose you saw a scarlet king snake that had red, yellow, and white bands encircling its body and an eastern diamondback rattlesnake that was solid white. Would you know that the king snake was an albino and the diamondback was not? The book *Reptile and Amphibian Variants: Colors, Patterns, and Scales* by H. Bernard Bechtel explains such color variation among snakes and other reptiles and amphibians. More than two hundred magnificent photographs, including subjects such as snakes, frogs, and salamanders, come as no surprise in a book about the causes of color variation. Many snakes have startling color patterns naturally. Add to these patterns the genetic anomalies that result because black, yellow, or red pigments have been altered, and some weird-looking animals appear. Some noteworthy biological messages also emerge.

A series of photographs of copperheads makes several special points. Most people are at least familiar with the name of this venomous snake, which characteristically has a tan to copper-colored head and a series of saddle-shaped crossbands down the body. Dark brown bands contrast against a lighter brown, coppery body. Coiled in the dead leaves of autumn, a copperhead blends in perfectly, virtually invisible to the untrained observer. Among the photographs of this master of camouflage are ones of copperheads with rare, genetically caused pigment properties that make them distinctive. One has stripes running longitudinally down the body instead of the normal condition of saddle-shaped, copper-colored transverse bands. Another is solid black, and another is tan with no markings. The albino copperhead is not white but has an orange body with reddish crossbands.

Why isn't it white? As explained in the book, true albinism in any animal is caused by a genetic condition that disrupts the metabolism of melanin, the black or brown pigments of animals. Melanophores are cells in the skin that manufacture melanin. If no melanin is produced by an animal's body, dark coloration will not be expressed. For most mammals, including humans, albinism results in a solid white skin and hair with the characteristic pink eyes. Many species of snakes and other animals normally have red and yellow pigments as well as black and brown. Snakes produce melanin as birds and mammals do, but unlike the warm-blooded vertebrates, the reds and yellows in snakes are produced by different cell types. Hence, when the dark melanin pigments are not expressed in a snake, the color pattern is produced by other cells. So, in an albino scarlet king snake or copperhead, only the dark color normally produced by the melanin cells disappears. The reds and yellows remain, sometimes being expressed more dramatically than when dark pigments are present.

Nonetheless, solid white rattlesnakes, cottonmouths, and alligators are known to exist—but these are not albinos. Instead they lack all pigment cells—except, for some as yet unknown reason, the eyes may be blue or black. This spectacular and extremely rare condition, known by the unfamiliar term *leucism,* can occur in any reptile or amphibian. Only one leucistic eastern diamondback, discovered in Florida by a road crew, has ever been found. The white alligators of the Audubon Park Zoo in New Orleans, taken from a single clutch in Louisiana, are leucistic.

Melanism is a condition that gives an animal an appearance opposite that of albinism or leucism because of an excess of the normal dark pigments. Thus, the color pattern is totally black. Melanistic salamanders, rattlesnakes, cottonmouths, and other snakes have been reported.

Reptile and Amphibian Variants is full of intriguing photographs of other color aberrations among reptiles and amphibians. A bullfrog that looks blue. A calico-patterned eastern garter snake, which normally has a black body and yellow stripes. A green Burmese python, a species that is characteristically brown and black. Perhaps the most important message of the many to be learned from color abnormalities is that genetic potential for extreme variability exists in nature. Hidden in the genes of any plant or animal on earth may be a phenomenon that could alter the course of medicine or agriculture, or produce yet another creature the likes of which we have never seen.

When Leaves Fall

Fortunately, for those who enjoy diversity, the natural world is full of color displays by plants and animals. Horticultural plants and domestic animals are less interesting ecologically because many have been bred to produce certain color patterns. But the colors of native species give a special message about the lifestyle of the organism. The explanation may not be obvious, but somewhere, in either the past or present, is an explanation for the color you see. So the next time you see a nocturnal moth with red or yellow wings, a brown chipmunk with white stripes, or a trumpet vine with orange flowers, try your hand at being an amateur ecologist. Ask yourself, Why is it this color instead of another?

You do not have to stick to single species to ask natural history questions about color. Autumn is a season of color. A visit to a classroom in any grammar school during autumn is sure to find bulletin boards full of paper cutouts of red, yellow, and orange leaves. A trip through the Great Smoky Mountains National Park on the right weekend can become an all-day affair as thousands of people admire the spectacle of variegated autumn trees. Even a walk through a local woods can offer a many-hued picture.

What is the explanation for autumn leaf colors? Why do some trees have brown leaves, others have brilliant red or yellow ones, while still others stay green all year? And why is it so difficult to predict exactly when fall colors will appear? Do ecologists really know?

Three basic pigments are responsible for most annual color patterns in plants. Chlorophyll, which makes the leaves of most plants green, is, of course, the dominant pigment. As days shorten in the fall, some leaves display two additional pigments. Carotenoids, believed important for capturing light energy, contain mainly yellows and oranges. Anthocyanins, involved in sugar storage, are red. The interaction and expression of the pigments that produce the variety of fall colors are at the mercy of the environment. Temperature, the amount of sunlight, and moisture conditions all affect the pigments and complicate ecologists' predictions regarding fall foliage. The timing and intensity of environmental events are critical in influencing the color of leaves. For example, cool autumn temperatures cause chlorophyll to degrade in many deciduous trees. Thus, the carotenoids and anthocyanins, normally masked by the chlorophyll, are accentuated, creating a colorful display of reds, yellows, and oranges. A sudden, heavy frost, on the other hand, may break down not only chlorophyll but also the accessory pigments. The result is muted colors. If autumn cooling is too gradual, colors may be dull because chlorophyll is still in the leaves, preventing full expression of the brightest yellows, reds, and oranges. To complicate predictions even further, the pigments respond differently to temperatures on the basis of when and how much rain fell during the past few days or weeks.

Although not fully understood, even the extent of summer rainfall can have an effect on fall foliage. In a wet growing season, plants produce more sugars and presumably more pigments. Therefore, if the chlorophyll is broken down rapidly, the remaining colorful pigments may be at their highest density and autumn colors will be at peak intensity. A wet summer could mean more and healthier leaves so that more leaf surface, and therefore more color, is actually displayed in the forest. The amount of sunlight also affects fall colors. Bright sunlight following several overcast days at the right season can result in a brilliant display of colors under certain environmental conditions. Despite what you may read about the best time to view the fall colors, no one knows for sure.

Finally, the geographic region and the types of trees themselves have an important influence on color. The brown autumn leaves characteristic of many trees, particularly in the warmer areas of the South, are usually indicative of the presence of pigments other than those already mentioned. Pigments associated with tannins, which mask the red and yellow pigments, are especially common. Some botanists believe the chemical makeup of tannins discourages insects from feeding on the leaves.

Evergreens, such as pines, firs, and magnolias, are another special case. The chlorophyll of evergreens has properties that allow it to withstand extreme cold during winter. As a result, other pigments are seldom expressed. However, in some situations even pine needles take on a golden appearance before they fall, thus contributing their part to autumn color.

The physiological function of plant pigments other than chlorophyll is poorly understood. Even less is understood about whether the display of color by a plant at the end of its growing season has any ecological function. Perhaps there isn't one. Although the purpose of the colors produced by leaf pigments remains unresolved, we can still marvel at the vivid display of trees during autumn, the season of color. The more we appreciate the complexities of nature, from all perspectives, the more likely we are to want to preserve and maintain the delicate networks that make up our natural world.

A Tadpole Tail

Our environmental networks are infinitely complex because each individual in nature interacts with and is affected by individuals of its own and other species. Every species is a kaleidoscopic reflection of its evolutionary past. Consider the cricket frog tadpole. Why do some of these small creatures, which have translucent tails, have black tips on the end, whereas others do not?

A few years ago, the British Museum of Natural History in London had a display called "Tadpoles—A Natural Division." The exhibit, part of a general section on the origin of species in the Darwin Hall of Evolution, was based on a study conducted by Janalee P. Caldwell on the U.S. Department of Energy's Savannah River Site in South Carolina. Caldwell's interest in the ecology of tadpoles originated in Kansas. She observed that cricket frog tadpoles from large lakes where fish were present were likely to have translucent tails with no markings. Tadpoles from small ponds without fish were likely to have black-tipped tails. When she began research at SREL, she observed the same phenomenon and wanted to explain it.

Cricket frog tadpoles are a favored food of fish and dragonfly larvae. Fish tend to catch swimming tadpoles and swallow them whole. To survive, a tadpole must be careful about swimming in open water. A totally translucent tail helps increase the chance of survival by making the small prey almost invisible as it swims through the water. However, even when fish are absent, tadpoles must worry about dragonfly larvae, equally fearsome predators to small aquatic creatures. Dragonfly larvae feed by grabbing the head of their prey as it rests on the bottom. A cricket frog tadpole, with its black head, makes an easy target for a hungry dragonfly larva. But a tadpole with a black tail tip can create confusion by appearing to have two heads. If the dragonfly grabs at the "wrong head," the tadpole is likely to escape. Ecologists call this phenomenon a deflection mechanism because it serves to deflect an attack away from the most vulnerable part of the body. A tadpole with a black-tipped tail has about a fifty-fifty chance of escaping from a head-grabbing predator.

So how was natural selection at work in evolving and maintaining such a color pattern? Although many natural phenomena are extremely complex, the cricket frog tadpole story appears to be relatively simple. Cricket frogs come from their terrestrial homes to aquatic habitats to breed. Because tail color is a genetically inherited characteristic, the parents produce tadpoles with completely translucent, black-tipped, or black-mottled tails, depending on their genetic background.

In a lake with fish, tadpoles without black-marked tails are best camouflaged and less likely to be eaten. In a pond with dragonfly larvae, those with black tail tips are most likely to survive. Those with black-mottled tails are susceptible as prey in either habitat. Froglets emerging from ponds with numerous dragonfly larvae will be mostly from black-tailed tadpoles, as the predators are most likely to have consumed those that had mottled or completely translucent tails. Froglets from lakes with fish are more likely to have no black on their tails. If the adult frogs return to the same lake or pond to breed, most of the offspring will have the color pattern with the best survival possibilities. The mottled forms are susceptible to predation in either situation, so why do they persist in most aquatic habitats? Why is this genetic line not eliminated completely? The explanation lies in the phenomenon of genetic crossbreeding, with black-mottled individuals being the result of a cross between the black-tipped form and the totally translucent form. Such crossbreeding can be common in natural situations because the levels of fish or dragonflies may vary unpredictably from year to year in aquatic breeding sites of cricket frogs.

Octopus Watching

An exciting feature of zoology is that animals are continually teaching us something new about themselves. Sometimes the surprise is that a particular species is capable of certain behavior. On other occasions, we discover that earlier notions about the behavior of a species were inaccurate. Along these lines, behavioral experiments have revealed previously undiscovered abilities of octopuses and have called into question a capability of dogs.

The experiments with octopuses were conducted by two Italian scientists, Graziano Firorito and Pietro Scotto. The results did more than simply reveal what these eight-tentacled creatures of the sea can do. They demonstrated a capability for observational learning, a previously unknown ability among invertebrates.

The investigators trained octopuses in laboratory aquariums to choose a ball of a particular color. This is an easy lesson for an octopus. Octopus eyes are similar in many respects to our own and can see colors and shapes. In the first part of the experiment, Firorito and Scotto presented octopuses with a red ball and a white ball. Except for the color, the two balls looked identical to an octopus. To teach an octopus to choose red, they gave it a fish when it put its tentacles around the red ball. When it grabbed the white ball, they gave the animal a mild electric shock for making the wrong choice. The investigators assigned red as the correct choice for one experimental group of octopuses and white for the other.

After an octopus had been trained to select either a red ball or a white ball (a process that only took a few trials before the "right" color was learned), another octopus was placed in an adjacent tank as an observer. The observer octopus was allowed to watch the conditioned octopus during four trials in which no errors were made. The observer octopuses were then tested in isolation. When given a choice of the two differently colored balls, they consistently selected the same color as the octopus they had been observing. Observational learning, characteristic of humans and some other vertebrates, is considered to be an early step in conceptual thought and the cognitive learning abilities of animals, the mental processes by which knowledge is acquired. The findings are especially intriguing because octopuses have highly developed invertebrate brains that have neural organizations similar to those of vertebrates.

A Puppy Dog Tale

In a test of a different sort, two researchers reported a finding that might cause some reconsideration about the use of dogs to identify the scent of suspected criminals. I. Lehr Brisbin of SREL and Steven N. Austad of Harvard University performed experiments on dogs that had already been trained to discriminate between scents from the hands of individual humans. The researchers' goal was to evaluate whether the dogs would distinguish between individuals when the scents came from a body part other than the hand.

In a series of trials, the dogs showed a high success rate in discriminating between objects touched by the hand of the trainer versus those touched by another person, a task for which they had been trained. In fact, this is a common test in advanced dog-obedience competition sanctioned by the American Kennel Club. But when the investigators transferred scent from the crook of the trainer's arm to the object, the dogs did not distinguish it from one touched by the stranger's hand. The scientists concluded that dogs indeed may have evolved highly sensitive powers of discriminating among scents. Such an ability could be important in the complex details of kin selection and recognition of subtle environmental cues. However, the findings have an aspect that potentially has immediate application to the lives of some people.

Frequently, law enforcement dogs make scent identifications of individuals on the basis of articles such as gloves, shoes, or articles of clothing, each of which would have had a scent imparted to them from different parts of the body. If, as shown by the research of Brisbin and Austad, such dogs do not make this discrimination automatically, could mistakes in identification occur? According to Brisbin, the lives of some death row inmates could actually hang in the balance on the basis of such evidence. The findings seem to call for reevaluation of the techniques by which dogs are trained and used to identify suspects in law enforcement.

From all areas of human endeavor, research can be found that helps lay the foundation for solving problems faced by humankind. An undeniably important issue, which will always be with us, is that we do not know everything about how the world works. Often, what we do know comes from basic ecological studies, which are intrinsically interesting because they explain the living phenomena around us.

5

Ever Smoke a Colorado Toad? On the Lighter Side of Ecology

I'm truly sorry man's dominion
Has broken Nature's social union.

So said Robert Burns, the Scottish poet, in his 1785 poem "To a Mouse." Burns had accidentally destroyed the mouse's nest, and he felt bad about it. Burns's poetic view of nature extended beyond the mouse in the field to all animals, plants, and habitats. Today many feel this way about natural ecosystems; we regret the ruin of "Nature's social union."

England's Romantic poets lamented lost loves and celebrated mighty battles. But much of their poetry was dedicated to an appreciation of nature. These environmental attitudes of two centuries ago are pertinent and applicable in today's world. John Keats, in *Robin Hood* (1818), sounded like a twenty-first-century environmentalist when he spoke of how Robin and Maid Marian would feel if they returned to Sherwood Forest.

She would weep, and he would craze;
He would swear, for all his oaks,
Fallen beneath the dockyard strokes,
Have rotted on the briny seas.

Will Americans someday echo this lament for the mighty trees of the Northwest forests that have also traveled on the high seas, not in the form of ships but on their way to Japan as commercial lumber or wood chips for making pulp?

Almost anyone who has ever taken a high school English class will remember the first line from Percy Bysshe Shelley's 1820 poem "To a Skylark": "Hail to thee, blithe spirit." Shelley goes on to say:

Teach me half the gladness
That thy brain must know,
Such harmonious madness
From my lips would flow
The world should listen then—as I am listening now.

The melody of songbirds, murmur of honeybees, splash of streams and rivers spoke to the Romantic poets—and they listened. William Wordsworth, in his 1805 poem also titled "To a Skylark," celebrated the sounds of nature when he wrote: "There is madness about thee, and joy divine / In that song of thine." Walk through your local woods, maybe even your own backyard, and listen to the brown thrashers, mockingbirds, and cardinals. You too can hear nature's message. You need but listen.

One perplexing attitude sometimes used to justify environmental destruction is based on religious grounds. Throughout history, human beings have used religious beliefs to justify their actions, including war, exploitation, and slavery. But the oft-quoted passage from the Bible that humans "have dominion . . . over every living thing" has more than one interpretation. One might be that people have a responsibility to take care of all the other creatures on Earth. We think a more appropriate interpretation is that people have a responsibility to live in association and concert with the rest of the living world. In any case, if one insists on a supremacy role for human beings, a conscientious head of household does not abuse the house, nor its inhabitants.

Samuel Taylor Coleridge's *Rime of the Ancient Mariner* (1797) offers sage counsel about the inherent worth of all God's creations:

> *He prayeth best, who loveth best*
> *All things both great and small;*
> *For the dear God who loveth us,*
> *He made and loveth all.*

Anyone inclined to engage in or support projects that result in the uncaring or even incidental destruction of native wildlife should reread this poem.

Coleridge's poem might be a fitting conclusion about attitudes of Romantic poets toward environmental preservation. But in his final stanza, Robert Burns says something to the mouse that may hold an even more important lesson for us.

> *Still thou art blest, compared wi' me!*
> *The present only toucheth thee;*
> *But och! I backward cast my e'e,*
> *On prospects drear!*
> *An' forward, though I canna see,*
> *I guess an' fear.*

Advertising Is a Natural Act

Ecology pervades our lives in a variety of ways—and always has. To develop a sense of the close personal connections between human cultures and the natural biological world, turn to the Yellow Pages in your phone book. How many services and products can you find that do not in some way mimic a plant or animal? You may find a few that are uniquely human, but not many.

For example, under the A's you will find Airlines, Air Conditioning, and Advertising. Each has its counterpart in the natural world. An airplane is simply a means of airborne transportation. For centuries plants have depended on the airline service of insects to transport pollen from one flower to another. Many shrubs and trees rely on birds to transport seeds to other areas. And of course millions of insects, birds, and bats use flying as a means of personal travel. Air conditioning is definitely not restricted to human use either. Honeybees will wave their wings in unison to fan a hive during extremely hot weather, lowering the temperature several degrees. The most widespread phenomenon of these three examples is advertising. In the natural world we find false advertising as well as advertising directed toward a particular audience. As with warning labels (and some of today's political messages), advertisement is often used to caution the audience.

Like human beings, other organisms make effective use of color and sound in their advertisements. One obvious use of color is that of brightly colored flowers. Their customers are insects, essential for pollination; the advertised product is nectar for the insect. Male birds, frogs, and katydids use sound to advertise to females their availability for mating. And a male lion or alligator loudly announces its presence to neighboring males, the message being Do Not Trespass.

Many animals rely on the chemicals known as pheromones to advertise their location to members of the opposite sex. Pheromones are used by female moths to attract males prior to mating, with each species producing a different pheromone. Bolas spiders, which eat moths, produce chemicals that mimic moth pheromones. To ensure a wide selection of mealtime choices, the spiders fill the night air with chemicals that include the critical ingredients of the pheromones of numerous moth species, a generic perfume guaranteed to attract male moths: the height of false advertising.

Such advertising scams are practiced by many species in search of a meal. A baby copperhead has a bright yellow tail that contrasts with the rest of its body camouflaged among fallen leaves. Upon seeing the slowly waving tail, small lizards or frogs foresee a quick meal in the form of a worm. Instead, they become a meal themselves for the unseen con artist. Or, equally as dramatic and dangerous is the open-mouthed display of the alligator snapping turtle with its lure of a pink tongue that looks like a worm ready for eating. Lying on a river bottom with jaws open, the alligator snapper beckons any passing fish to drop in for an easy meal.

Humans rely heavily on lights for advertising, as do some animals such as the lightning bugs seen in backyards across the United States. The male fireflies blink their lights in a code that indicates they are interested in and available for courtship. The female firefly watches the night sky from her location on the ground or in vegetation, and with a light of her own answers the male's signal. Many species of lightning bugs may be active at the same time and place, especially in the tropics. The fireflies obey that well-known advertising maxim Know Your Audience. So the blinking codes characteristic of the different species vary, thus preventing mating mix-ups.

False advertising is common among lightning bugs, especially in tropical situations. Such deception is practiced by the females and the luminescent larvae of species that eat other species of fireflies. Upon spotting a flying male of another species the imposter responds by delivering coded flashes. Rather than flashing the code for its own species, the ground-based mimic gives the signal characteristic of a female of the same species as the male. The male flies down, ready to court, but unexpectedly becomes a victim of a system without truth-in-advertising laws.

You have only to read the label of any medicine, cleaning agent, or appliance to realize that warning advertisements are widespread these days. Such notices are especially common among animals. The tail vibration of a disturbed rattlesnake creates the loud whirring sound that cautions an intruder. The bright red coloration of an eft (the terrestrial stage of a salamander known as the eastern newt) serves as a warning to birds, which can see color, that it is poisonous to eat. A snarling bobcat or raccoon sends a pretty clear message that it should be avoided.

Returning to the game with the phone book, you can play it with any part of the Yellow Pages. Most human products and services are duplicated in nature. But you really do not have to go beyond the A section to find a heading that indeed is uniquely human: Attorneys. Although some do make good friends and relatives, apparently nothing comparable exists among other animals and plants. As far as we can tell, lawyers are not a natural phenomenon.

Who's Keeping Score?

Like phone books, newspapers can provide insight and perspective about human relationships with the natural environment. But the message found in newspapers does not necessarily reflect our true connection with the environment. For example, millions of people check the sports pages of newspapers each day for updates in the world of athletics. How many check the environmental pages for the status of life on Earth? Easy answer, None. Newspapers do not have a daily environmental section. Why not? One reason is that sportswriters create hype about sporting events in a most effective manner. Ecologists have not yet developed comparable techniques for focusing public attention on the environment.

One distinction between sports and wildlife trends is the length of the season. All sports begin and end in less than a year—although sometimes it seems longer—and the hype of highlighted games seldom lasts more than a few days. The decline of a plant or animal species is measured over years and decades, over generations and lifetimes. The chains of environmental events are too long for our attention spans. People cannot be expected to keep track of the slow but inexorable loss of thousands of species as tropical rain forests are diminished, oceans are polluted, wetlands are destroyed. Often we only hear about the final outcome for a species when it is already too late to have a winning season.

In contrast, some sports receive print and broadcast media exposure (or overexposure) on daily and weekly victories and failures throughout a season. And the number of teams involved is small relative to the number of species in the natural world. Continual sports updates presage the Super Bowl, World Series, NBA Championship, and Stanley Cup Finals. By the time the culminating event is reached, millions have followed the contests and anxiously await the outcome. For wildlife, however, we often emphasize only the climactic event, extinction. Severe declines in biodiversity (struggling through the playoffs) and disappearance of a species (losing the final game) are noteworthy to some. But most people are uninformed or uninterested. Minimal attention has been given to the numerous events and statistics that led to the showdown.

Reasons for the lack of extended interest in the fate of wildlife or natural habitats are several. You can check the win-loss record for any sports team and whom they play next. But you cannot find the change in status of endangered species, even those with the same names we find in sports. Do we have more or fewer timber wolves in Minnesota or panthers in Florida than last year? Is the next competitor for each species a new highway, a golf course, an automobile factory?

Such environmental information, when available, is usually restricted to special newsletters, scientific journals, or wildlife reports. The environmental playoff may warrant a few articles in the newspaper, but the presentation is not a neat package. In contrast, anyone can read a newspaper to find out not only that the Atlanta Braves beat the Houston Astros on 7 July 1996 but that Greg Maddux was the winning pitcher (with a season record of nine wins and six losses) and that Fred McGriff hit his twentieth homer for the season. Sportswriters provide statistics with no qualifying disclaimers, and they are available throughout every week of every year.

Ecologists do not post weekly box scores for declining species. Most scientists are cautious about what they present as fact, and determining the population status or basic ecology of a species is an uncertain process. Ecologists can only make estimates that reveal general trends of declines or increases. They cannot provide the clear-cut, concise statistics on a daily basis that sportswriters can. Yet, though the wildlife scores are not available, the games are still being played.

Perhaps if records of the decline of native species were presented in a consistent fashion, people would be more inclined to follow environmental trends. Being more aware, many would become fans and get involved in influencing the outcome. We may never have environmental pages that rival the sports pages of newspapers, but if we did we might be able to avoid some of the environmental playoffs we are headed toward—with no winners and no chance of the losers ever making a comeback.

The Ecology of Dragons

A close environmental accounting would show that most wildlife species are being reduced in numbers worldwide. The statistics may not be as exact as for sports, but the circumstantial evidence is clear: More and more species that would have been content to be cellar dwellers are now out of the game forever. Dragons never made an endangered species list, but a few centuries ago dragons would have been prime candidates. Their rarity today is ample evidence, although it might be premature to suggest they are extinct. We know they existed back then because everybody said so, particularly those who were educated and could write. But what do we really know about the ecology of dragons?

People do not ordinarily make something up without some basis in fact, so what was the origin of dragon lore? Modern-day reptiles that are typecast for the role of dragons are the crocodilians. These big reptiles are very private creatures that do not willingly divulge their ecological secrets, even to scientists. Plus, even modern-day reptiles are often given credit for behavior far beyond their abilities. Promoting a little imagination and superstition among an uneducated and gullible population a few centuries ago would be no more difficult than today.

The local name for the Chinese alligator, a creature not unlike the American alligator, was "earth dragon." Even today's scientists have a tendency to fill in the biological blanks when they do not know the answers, so it is easy to see how some imaginary dragon feats developed. Reports of an airborne crocodile that acted like a flame thrower went unchallenged. Judging from what people seem willing to believe about sightings of Big Foot or the Loch Ness Monster, the early perpetrators and believers of dragon myths have a lot of descendants. A form of reinforcement that monsters like dragons existed came eventually in the form of dinosaur fossils found throughout China. Big bones meant big animals, and the dragon lore fit in perfectly. After all, is a Chinese dragon any more improbable than a *Tyrannosaurus rex*?

Among the Chinese ecological observations about dragons were that their favorite bird was not chicken or turkey. Instead, they loved to eat swallows. In fact, people who had eaten swallows were advised not to swim rivers lest they become a meal for a river dragon. One can only wonder how frequently people swam rivers after a meal of swallows. Perhaps most environmentally important is that dragons were credited with controlling the weather, especially rainfall. Thus they were ultimately responsible for floods and droughts. Early Babylonian lawyers probably debated cases about whether a person with dragon insurance could claim flood damage to his house.

Once dragons had settled down in the Far East, they behaved with dignity and were held in awe. Eventually, they began to migrate westward and became more quarrelsome toward humans, engaging in fights with saints, ocean travelers, half-gods, hobbits, and a few regular guys. Once they had swum or flown across the English Channel they had acquired a taste for beautiful virgins, delighted in opportunities to take on regional heroes in mortal combat, and were forever causing problems to villages and kings. But all animals evolve over time. It is the least we would expect from the mighty dragons.

Peter and the Ecology of the Wolf

Presumably the ancients never held protest marches to protect dragons. But today seemingly anything can become an environmental issue, even "Peter and the Wolf," "Little Red Riding Hood," and "The Three Little Pigs." Some time ago, a news channel presented a human interest story about environmentalists boycotting a concert of "Peter and the Wolf." One member of the orchestra (I think it was the cellist) even resigned rather than participate in the concert. The boycotters felt that portraying the wolf as the bad guy turns people against the environmental protection of wolves. I thought it was a joke. (Before I go on, let me admit that in my book *Keeping All the Pieces,* I mentioned that such stories probably had not helped attitudes toward wolves. But I would never suggest we stop reading the stories, because I think children's attitudes are tempered by whoever is doing the reading.)

I find the censorship of "Peter and the Wolf" hard to take. For one reason, the argument itself is fallacious. Why would anyone be against the environmental protection of wolves because a wolf ate Peter's duck? (I always wished he would eat that little bird that sounds like a flute.) "Peter and the Wolf" was my favorite humming tune when I was a little boy. In fact, when I listened to the story, I remember feeling bad when the bass drum announced the arrival of the hunters who would dispense with the wolf and save Peter. I never particularly liked Peter. Could this be the reason I do not care much for string quartets, which represented him? I also wanted the wolf to eat Ms. Hood as well as all three little pigs instead of just two. Of course, in more modern versions the wolf doesn't eat any pigs;

he merely chases them to the next house. The point is that many people heard such tales and yet still grew up to appreciate wolves. Let's not get too carried away in trying to make some of our classics politically correct. Expressing concern for the environment and encouraging people to respect wildlife is fine. Boycotting "Peter and the Wolf" seems comparable to censoring *Huckleberry Finn* or banning unexpurgated editions of Shakespeare. Apparently, some environmentalists need to learn to separate art from ecology. Such protests help neither the environment nor the wolf. Carried to extremes, such attitudes might lead to changing the original story so the wolf and Peter discover they are long-lost brothers.

Another little tale about a persecuted species was told to me by Curt Richardson of Duke University. Curt has a research site in the Everglades where he studies the effects of phosphorous on the ecosystem. He takes teams of students on an airboat to floating platforms miles from shore where they sample the vegetation. Parts of the ecosystem that he does not collect, but appreciates nonetheless, are the big cottonmouths that coil up and bask on his platforms as they would on a log. When the investigators arrived to sample on one trip, Curt was pleased to see a particularly enormous cottonmouth that is often present on one of the platforms. He enjoys showing the students what a big venomous snake looks like.

While he tied up the airboat, some of the new students went on ahead to the platform. When he arrived, the snake was gone. "Where's the cottonmouth?" he asked, knowing that it was seldom alarmed by the presence of the researchers and had been there minutes before. The students proudly proclaimed that they had driven it away for safety reasons. At least the students did not kill the snake unnecessarily.

Curt explained that although cottonmouths will stand

their ground, they are not aggressive, and a person will have no problem with one that is left alone. Where had the snake gone? Into the sawgrass, out of sight, "right over there." Curt then delivered the news that the sawgrass "right over there" was the sampling area where *they* now had to go. An unhappy cottonmouth was now hiding somewhere in the vegetation they would be sampling, instead of being safely coiled on the platform in full view. Perhaps the overriding message in the wolf and cottonmouth stories is that we are often better off if we simply leave art and nature alone and accept them the way they are.

Smoking Toads

The interaction among environmental issues can sometimes lead to fascinating situations. The perceived decline of amphibians worldwide is clearly an environmental concern. Smoking is less clearly so, though some people seem bent on making it one. But a combination of the two concerns, Colorado river toad smoking, gives a new perspective to both issues. Toad smoking has presented a problem that must be addressed by fish and game officials as well as narcotics agents.

How does one smoke a toad? Toads are smoked like cigarettes or cigars, not like oysters or salmon. The narcotics aspect started with toad licking, in which one laps up the toxic secretions on a toad's head to achieve psychedelic effects. But by consuming toxins, that is, poisonous substances, a frequent licker can become ill, maybe even die. But do not dispair. Clever Californians have found a way around this problem. Heat breaks down the toxic components in the toad's glandular secretions without affecting the sought-after hallucinogenic compounds. The secretions can be dried and rolled in cigarette paper, and the smoker is ready for a trip. Because the enormous Colorado River toads are under assault by California toad smokers, concern has been raised about their (the toads') environmental welfare. A positive outcome from all this might be antismokers becoming a little more tolerant of cigarettes and cigars upon discovering that some restaurants are considering putting in a toad-smoking section.

Such complex issues as whether the Romantic poets were environmentalists, whether children's stories can induce anti-environment attitudes, or whether narcotics or wildlife agents should deal with hallucinogenic amphibian lickers are trivial compared to the universal environmental question: What will the weather be tomorrow and next week?

Rolling the Weather Dice

Ecologists have at least one thing in common with other people: a fascination with weather. The question of how plants and animals respond to rainfall and temperature dominates many environmental studies conducted by research ecologists. Concern for these particular environmental variables seems to be important for many individuals who are not ecologists.

Human beings can discover a lot from plants and animals about a variety of subjects, and weather prediction is one area where we can learn a valuable lesson. The lesson is that day-to-day, much less week-to-week, variation in the weather cannot be consistently and reliably predicted by organisms, including people. Of the millions of different species, we know of none that bases its behavior, physiology, or other biological plans on a five-day weather forecast. Among wild animals, any that based their survival on an assumption that they were able to predict the weather eventually left no descendants. That kind of thinking does not exist in the animal kingdom, except among humans. Despite the weekly confirmation that a long-range weather forecast is absolutely meaningless, we still faithfully check to see what weather is supposedly in store.

We have the impression that some animals can predict the weather; actually they are only responding to current meteorological conditions. For example, frogs sometimes call before a rain. However, high humidity, falling barometric pressure, and dark clouds are detectable even by human beings. The frogs are merely making a short-range prediction of a few minutes or hours. People could do that just as well. And sometimes both frogs and people are wrong about predicting rain within the next hour.

Birds that migrate north in the spring and south in the fall are not making long-range weather predictions either. Birds, like any of us, can predict with 100 percent assurance that winter in Michigan will be colder than summer. Predicting seasonal changes is something that all surviving species of plants and animals can and must do. In the north temperate zone, plants and animals have evolved to prepare for the change in the seasons. Many plants, such as tulips and fruit trees, depend on an extended period of cold temperatures before they will bloom. Yet no organism is able to predict the weather of next week or even of tomorrow.

One perplexing type of weather prediction is that associated with global warming. Some records suggest that our mistreatment of the world's atmosphere is causing a worldwide temperature increase, but results are not yet conclusive. A major culprit of atmospheric warming is carbon dioxide emissions from automobiles and other burners of fossil fuels. The prospects are indeed alarming, but long-term temperature data to support the claim are confounded by the great variability in local as well as global weather patterns. One of the complications is volcanic eruptions, such as Mount Pinatubo, which produced atmospheric gasses that actually resulted in a slight cooling of the earth's temperature.

The vagaries of long-range weather reporting were confirmed in a science fair project undertaken by a young relative of ours. She kept a record of the five-day weather forecast published in the newspaper, with regard to whether it would rain. Each day she then recorded what the actual weather was—did it rain or not? We were impressed. The weather forecast was right almost as many times as it was wrong. You could flip a coin and your chances would be just as good at predicting whether it would rain or not five days later. However, we are not so naive as to think weather stations that provide these five-day reports actually flip a coin. We think they probably use a pair of dice. That way more possibilities are available. After all, they provide us with the exact high and low temperatures that we can expect.

Meteorological studies have shown that one of the safest overall weather predictions is that tomorrow's weather will be like today's. Obviously your prediction will be wrong much of the time but no more so than standard weather forecasts. So why do we continue to pay any attention to a weather forecast that is more than a few hours away? We don't know the answer to that question. But we have a prediction to make: Tomorrow you will check the weather report in the newspaper so you can plan for the rest of the week. So will we.

6

Barbershop Ecology: True Tales Worth Telling

Early in 1994 I caught my first python in the wild. The seven-foot-long constrictor was one of twenty-five we caught. We released them at the site of capture, as our objective was simply to see the habitat, observe snake behavior, and become familiar with a new group of animals. "We" were a group of American, Japanese, and Swedish ecologists hosted by Rick Shine and Thomas Madsen of the University of Sydney on a field trip to the Northern Territory in Australia.

In Australia everything is backward, except the people. On a summer day in January we went *up* toward the equator, driving in the left lane, to a tropical area near a town called Humpty Doo. I became used to the constellation Orion being upside down and the dog star, Sirius, being on the right instead of the left.

Even for a herpetologist, snake hunting in Australia is something of an adventure. For one thing, most snake species belong to the cobra family. (In the United States, only the coral snake belongs to that group.) Many of the snakes in Australia are considered to be "deadly poisonous." Rick Shine told us that the first rule to follow with an Australian snake is "don't pick it up." The second rule—if you're a herpetologist with split-second reactions and you're certain that it is one of the few nonvenomous species—is "catch it by hand." The venomous ones must be handled with special techniques.

Observing nature in a totally different environment can provide ecological perspectives about the way our world works. One night on a collecting trip, as we traveled by boat through mangrove swamps near the Arafura Sea, Steve Arnold of the University of Chicago and I observed an unusual phenomenon. At least it would be unusual in North American river swamps. Harmless mangrove snakes lay stretched out on mud flats near the water, and they were extremely easy to catch. Few tried to escape. Likewise, the crabs and the mud skippers (fish that walk around on land) showed no alarm at our presence. Why would animals let a potential predator approach them so closely?

No one knows the answer for sure, but the shores of southern U.S. rivers, where we collect ever-wary water snakes, are patrolled by fierce (to a snake) land predators: raccoons. A snake, crab, or land-walking fish that did not flee immediately along the shoreline of our rivers would be quickly eliminated. Their absence Down Under suggested to us that raccoons are a keystone species of North America—a species that controls the character of an ecological system.

Although the mangrove snakes did not display any defensive behavior, other animals did. One day I approached a blue-tongued skink, an enormous lizard nearly two feet long and as big around as a coffee cup. Rather than trying to bite me, it opened its mouth and displayed its huge tongue, which was as blue as the sky. I was not deterred from capturing it, but a predatory bird might think twice before pouncing on such a creature.

Many animals also have unusual methods for obtaining food. One is the death adder, aptly named because most people who receive a solid bite are unlikely to live. We found more than a dozen at night in the floodplain of the Adelaide River. Although a member of the cobra family, death adders have evolved to look like the unrelated vipers: they have fat bodies, elliptical pupils, and broad heads, not unlike our cottonmouths. A hungry death adder sits and waits for a frog or small mammal to come by. Then it lifts its tail, which is black above and yellow below, and begins to wiggle the end above its head. Seeing what appears to be an easy meal of fresh caterpillar, the prey animal captures the squirming morsel. The end is quick. The death adder's venom kills the other animal before it realizes it has become prey instead of predator. Interestingly, baby cottonmouths and copperheads in North America also have bright yellow tails that are used to lure small prey.

The opportunity to explore a new ecosystem, the Australian tropics, could not be passed up (even though I missed New Year's Eve altogether, leaving Los Angeles on 30 December and arriving in Australia on 1 January). But a week of four-foot-long monitor lizards, man-eating crocodiles, and countless snakes did not diminish my wonder at the everyday creatures that inhabit my backyard. Wherever you are, exploring nature is the greatest adventure, the final frontier to which anyone can journey.

More Than a Haircut

To hear about great nature adventures, all you have to do is get a haircut. Back in the United States I was pleased to be reminded that old-fashioned barbershops are still a good place to swap interesting information. Besides the usual theories of who will win the pennant and where the best local fishing spots can be found, one can pick up new ecological insights.

Russ Rader of SREL and I had been teaching summer courses at Michigan State University's Kellogg Biological Station when haircut time came around. Hickory Corners, Michigan, is not a very big place, but it does boast a barbershop in the vicinity. When we went for haircuts, Russ was wearing a T-shirt with a bug pictured on the front; I was wearing one with a salamander. Before we had even taken seats to wait, the question of whether we were ecologists from the local biological station was asked by one of the customers. After we admitted that we taught ecology, the men in the room began to reward us with some of their own ecological observations.

I was interested to hear that rattlesnakes are "all over the place" in the area. Based on my own observations and those of other herpetologists, the single species of rattlesnake found in Michigan is not particularly common. In fact, the hardest part of studying the massasauga, as the small midwestern rattler and only venomous snake in Michigan is called, is finding any. Apparently, like many of the snakes I hear about in other regions, Michigan rattlesnakes know to hide when a herpetologist is in the vicinity.

Someone asked what we knew about hummingbirds hitchhiking on the backs of geese. Russ quietly rolled his eyes. But I could not resist going along with what I thought was a tall tale. I remarked that I had wondered about those tiny thumbs on the right wings of hummingbirds. But the man was serious. He swore that hummingbirds make their long trek from northern states to the tropics by attaching themselves to Canada geese—free rides south for the winter, north for the summer.

Other tales circulated while the haircutting continued, and customers with fresh haircuts remained in the barbershop. During my turn in the barber's chair, six perfectly groomed men sat around swapping ecology stories with each other and us. An older man, who hadn't said much, stood up during a pause in the conversation and walked over to where I sat. Standing beside one of my sideburns while the barber busied himself with the other, he told me that as a child he had been trained by an Apache Indian to stalk deer and other wildlife. Now, even I can recognize a deer track and the direction the deer was walking, so I wasn't sure why I should be impressed. But there was more. After finding the deer, he could walk up and place his hand on it before it would run. He explained that anyone could be trained to approach wild animals without frightening them. Most of the technique involves mental attitudes, such as taking on the persona of another animal. The stalker must also use peripheral vision, never looking directly at his prey.

By the time this story was finished, so was my haircut, and Russ and I left. Later that day, we saw my friend Joe Johnson, a wildlife biologist. I told him about the hummingbird hitchhikers and the Apache stalking technique. He had heard of the Apache stalking technique but did not know anyone who could do it. To our surprise, Joe knew hunters who claimed to have seen hummingbirds leave the bodies of Canada geese that had been shot. He did not go so far as to say he had observed the phenomenon or even believed it was true. In fact, he pointed out that both the timing and destination would be imperfect.

Hummingbirds leave Michigan on their southern migration in August, with a few dawdlers staying around until October. Most of the Canada geese do not even arrive in southern Michigan until mid-September, staying around for several weeks and leaving in late October. To hitch a ride on a goose, hummingbirds would have to wait around longer than most do. Even then, they would only make it as far as Alabama, the final stopover point for most of Michigan's Canada geese. The final destination for a ruby-throated hummingbird during winter is Mexico or farther south.

What ecologists already know of the behavior and ecology of animals is fascinating beyond measure, and many bizarre behaviors are yet to be discovered. So we must be cautious in our judgment about unusual tales. Nonetheless, without more evidence, none of us can accept the theory of hitchhiking hummingbirds. But the Apache wildlife stalking story cannot be summarily dismissed, at least by me. Later that day, Russ and I saw a small rabbit sitting on the side of the road. I stopped the car and told Russ to wait while I stalked it, Apache-style. Pretending I was a large grazing animal like a buffalo or deer, I sauntered toward the rabbit. I looked across the road as if interested in some distant grazing pasture. Sure enough, the rabbit did not run until I finally stopped two feet from it.

Frogloggers Tell Many Tales

Perhaps, within the vast scope of modern human knowledge, ancient ecological wisdom awaits our discovery. Nonetheless, an association between technology and the environment has also had its positive returns. But have we become too technological in our attitudes, too reliant on mechanical solutions, forgetting some of the basic lessons our ancestors knew? The positive, such as air conditioning, cars, and hydroelectric plants, can also have negative returns: ozone depletion, smog, dammed rivers. For technology with only positives, try frogloggers, burrow cameras, and the discovery of rare frogs and turtles.

A froglogger is an automated recording system, a tape recorder designed to start and stop automatically when no one is around to operate it. Many birds, insects, frogs, and mammals communicate by sound. So a well-placed recording system can keep track of numerous forest sounds when no human being is there.

Some of our technology seems simple and is taken for granted once all the bugs have been worked out and someone shows us which buttons to push. Such appears to be the case with the froglogger after work by Charles Peterson of Idaho State University and Mike Dorcas of SREL. They tinkered and toyed with automated recording systems for purposes of detecting frog calls. The product is a recording device housed in a weather-resistant box and programmed to record for intervals from a few seconds up to a minute or more each hour. An internal clock, audible to the listener, announces the time when each recording begins so that ecologists can know when certain sounds were made. A solar panel keeps the battery charged, and the recorder can go for a week or more before tapes must be changed. Then begins the real fun with a froglogger. The tapes must be listened to and the sounds identified.

At a froglogger "party," six of us sat in my living room listening to the stereo play sounds of the night. The froglogger, programmed to record for a one-minute interval each hour from dusk till dawn, had been placed alongside a natural wetland at the SRS in South Carolina. Calling frogs can be detected in this way because frog choruses often take place over a period of several hours. With the froglogger tapes, we were able to compress two full weeks of nighttime frog calls into a comfortable three-hour session without getting rained on, without being bitten by mosquitoes, and without staying up too late too often.

We settled back to listen. Night in the forest away from city sounds is full of music, and we recognized most of the songs we heard. The rapid chuckling of leopard frogs and the clicking of cricket frogs were heard on every night. On cool nights we heard the cheeps of spring peepers, tiny tree frogs of the eastern United States; on warmer nights, the banjolike strums of bronze frogs. Once we heard a sound that reminded us of the dawn age of modern technology: a distant, mournful train whistle. Two barred owls challenged each other with their "who cooks for you all?" hoots, silencing the eerie warble of a screech owl. During a windy night we heard a distinct crash that we confirmed the next day to have been the fall of a dead pine tree that had stood near the froglogger station.

We may never know the identity of other sounds on the tape. Did we really hear a coyote howling? Was the large animal splashing through shallow water at the edge of the wetland a deer, a wild pig, an alligator? The most significant sound, and one we could all identify, was not unlike an old man snoring—the mating call of the gopher frog, one of the rarest amphibians in the eastern United States. The call was recorded on two tape segments during a rainy night. Gopher frogs are so rare they are known to be in only a handful of locations in South Carolina. In some years during the last two decades none have been seen or heard anywhere in the state for the entire year by the herpetologists who look and listen for them. The froglogger had confirmed the gopher frog's occurrence at a wetland where it was formerly unknown.

Frogloggers are destined to become a major environmental sampling tool. To elevate the technology further, Mike Dorcas is working with Ontario Hydro Technologies, a company developing computer software to identify frog calls. For example, a tape from a froglogger can be digitized into a computer sound file. The file can then be played into the software program for the computer to identify the calling species of frogs. The software can also eliminate sounds, such as passing trains or traffic, that might make it more difficult to hear frogs calling. The more rapid interpretation of sound tapes will result in a technology that will be of enormous value in environmental assessments of vocal animals.

No matter what new knowledge is in store for us about calling frogs, the froglogger has already answered one question that has nagged people for centuries. A tree falling in the woods *does* make a sound, whether or not anyone is there to hear it.

Exploring with a froglogger is an auditory adventure. Exploring with a burrow camera is a visual one . . .

I could feel the anticipatory tingle in the others as we watched the tiny TV screen, no bigger than a deck of playing cards. Everyone was silent as we peered at the eerie light and saw the walls of the tunnel go by, as Russ Bodie and Tracey Tuberville took turns snaking the camera cable along the winding path. The image on the screen showed a turn in the tunnel and beyond that a long, straight stretch. At the far end a dark form came into view.

Someone said in a soft voice, "What's that? Up ahead, at the end of the tunnel."

No one spoke for a moment as the form began to enlarge on the TV screen; then someone else said, "It's a tortoise! Look, it's moving."

A cheer went up from beneath the blue plastic tarpaulin we had used to shield ourselves and the equipment from the sunlight. All ten of us spoke at once, pointing to the image on the screen. The burrow camera had worked—we had found a gopher tortoise in its underground tunnel. Gopher tortoises are impressive in their ability to dig their own burrows, but we already knew tortoises could do that. We didn't know for sure that the burrow camera would be effective at finding them there.

A burrow camera is a simple device in terms of modern

technology, but its value for studying the secret lairs of animals that live beneath the ground is immeasurable. The device itself consists of thirty feet of coaxial cable (comparable to cable TV wire) about the size of a garden hose with a camera on one end and a monitor for viewing on the other. The camera is equipped with tiny infrared light sources so that the natural darkness in the deep tunnel appears fully illuminated, but only to the eye of the camera.

Many animals cannot see infrared light and may be unaware that their privacy has been invaded. People cannot see infrared either, but the camera can. The reflected light is converted into a picture visible to us in the form of images appearing on a TV monitor. A dark-as-a-cave burrow can look like a brightly lit room. We found the tortoise in the fourth burrow we examined with the camera at an active gopher tortoise colony in South Carolina.

These imposing turtles are officially protected in most re-

gions of the Southeast where they still occur. Gopher tortoises are known as a keystone species, a plant or animal that controls the character of an ecological system. Keystone species can dramatically alter the structure and dynamics of a habitat in direct and indirect ways. Gopher tortoises do so by modifying the landscape in the sandy soil habitats where their burrows serve as refuges not only for themselves but for many other animals.

A burrow may be more than thirty feet long and as wide and high as a full-grown gopher tortoise. The domed shells of big ones are more than fourteen inches long, ten inches wide, and eight inches high, and their hind feet look like they should belong to a miniature elephant. Gopher tortoises are the most terrestrial of the eastern turtles, living a peaceful life grazing on grasses and other vegetation and spending the off hours of darkness or cold weather underground.

Our excitement at seeing the gopher tortoise went beyond the thrill of using a piece of military-like technology that allows one to see in the dark. We were jubilant because this was the first confirmed sighting of a live tortoise at a newly discovered longleaf pine-wiregrass habitat, a rarity in itself. So unusual, in fact, is the habitat and its colony of tortoises that, according to Stephen H. Bennett with the South Carolina Department of Natural Resources, arrangements have been made to assure that a conservation organization, the Heritage Trust Program of South Carolina, will own the land and preserve the system.

The colony was special to us because it represented the northernmost one known to exist for this fast-disappearing species. But perhaps the greatest thrill of all, from an environmental standpoint, was that the newfound colony differed in another way from others reported in past decades. Almost without exception, a report about a gopher tortoise colony has

been one of disruption or local extinction from such causes as development, agriculture, highway construction, and the capture of individuals for pets or food. Despite official protection, colonies still are gradually giving way to the expanding human population, especially in Florida.

How could such a discovery even be made in today's world? How could a colony of animals the size of volleyballs go unseen by passersby, or a longleaf pine–wiregrass habitat go unaltered? I like the answer. The land had been owned, and protected from outsiders, for several generations by bootleggers. Their short-term impact on people was special, but their long-term impact on environmental preservation was even more so.

Happy Birthday—Here's Your Rattlesnake

Technological advances such as frogloggers and burrow cameras that contribute to environmental knowledge may help dispel some people's misconception that all ecologists and environmentalists are against technology. More than a few people still consider ecologists to be solemn, dour individuals with a doomsday attitude about today's world. Clearly, the Earth faces some problems that must be addressed. Ecologists and environmentalists often have a vivid idea of just how serious those problems are, and they have a somber way of discussing them. But every discipline has its lighter side, and it is probably a good idea to consider that side from time to time. Saving the world doesn't have to be a humorless vocation, and an appreciation of the natural wonders that still abound on Earth has its own rewards. Consider, for example, the nearly limitless gift ideas the environment offers.

Who would not want to hear, "Happy birthday, here's your rattlesnake"? Actually, here are your twenty-eight rattlesnakes, plus a handful of other kinds of snakes for good luck. Sound like a good time, or what? Well, that is the gift I gave my son, Mike, for his twenty-third birthday. An ecology field trip. The present was my idea, and I got to deliver it. His mother actually had in mind the kind of gifts you wrap up in paper and ribbon.

The presentation day was a Sunday in October at a national wildlife refuge, not an exotic pet store, near Orlando, Florida. Terry Farrell and Peter May of Stetson University let us go with them to their pygmy rattlesnake study site. These two ecologists have one of the most impressive snake studies in the country. They have collected more pygmy rattlesnakes than any other two people anywhere. Upon being invited to join them on a sampling day, I had asked if Mike could come along.

Now, Mike grew up like most other children, enjoying the outdoors and always wanting to catch something. Because I did a lot of fieldwork, he often went with me. For one of his birthdays we visited Laura Brandt in the Everglades during her study of American crocodiles. On another we went to visit Dick Vogt and Miriam Benabib at the biological station, Los Tuxtlas, situated in a tropical forest in Veracruz, Mexico, where he caught his first boa constrictor. When Mike yelled that he had found a big boa, which turned out to be longer than Mike was tall, Tony Mills, who had come with us from SREL, and I were both alarmed. We ran as fast as we could to where he was, trying to get there before he got too close to the snake. Our concern was justified, because we were near a place where, the year before, David Scott of SREL and I had caught one of the large venomous pit vipers some call a fer-de-lance. As it turned out, Mike had caught a boa, and Tony and I were mostly dismayed because he, and not we, got to catch it. Mike's sister Susan Lane, who had also gone on the trip, did not seem disappointed that Mike, not she, found the snake.

Mike also had birthdays in Arizona collecting snakes, and he celebrated others in Michigan and on the Potomac River capturing turtles. On some birthdays we just went to the local woods, river, or swamp to see what we could find.

For the pygmy rattlesnake hunt we met the others at a gravel parking lot alongside a swamp. As we drove up, one of the students was taking measurements on a pygmy rattlesnake that had been crawling across the parking lot as she got out of the car. We walked down a dike separating two lakes toward a hammock where most of the snakes are found. A hammock is a wooded area surrounded by open wetlands and with palmettos, cypress, and bay magnolia trees. A wide variety of wildlife find refuge in hammocks. As we walked along, Terry Farrell found a baby pygmy rattler coiled on the dike. Mike found another, coiled in the grass a few feet away. I relaxed. At least

Mike had found a snake; he would not be disappointed. The pygmy rattlers are so well camouflaged that many people do not find one on their first trip. On a previous trip with Peter and Terry, Tony Mills and I both had stared at the ground while someone pointed to a pygmy rattlesnake that neither of us could see at first. It blended perfectly with the ground cover of dead twigs and pine cones.

These smallest of venomous snakes in North America are shy and retiring when it comes to people. They prefer not to be seen or heard. They do have a tiny set of rattles that sound like an insect buzzing if they get mad and you get close enough. And their venom is potent, drop for drop. But because of their small size (seldom over eighteen inches), they are not as dangerous as cottonmouths or diamondback rattlers.

Mike was in tune with his environment that day and spotted another adult pygmy rattler on the dike. I began to wonder how many I had stepped on. Then came the really fun part, entering the hammock where most of the snakes lived. A walking trail leads through the hammock, but central Florida had recently been favored with tremendous rains. The trail was a foot deep in water, which was over two feet deep in the woods themselves. Some snakes seek refuge on higher ground or vegetation during floods. And the little rattlers were on exposed areas of dry ground as well as on palmetto fronds and in low-lying bushes. Mike found one more than six feet above the water in a wax myrtle tree. This discovery gave me a new regard for pushing vegetation aside while charging through the woods off the trail.

We stayed approximately three hours and found twenty-eight pygmy rattlesnakes. I say "we" found them because I was there. Mike found five, Peter found six, Terry found seven, the students found the rest, and I found none. But I did catch a green snake no one else saw. We also found two garter snakes, four ribbon snakes, a water snake, and four box turtles. Ecology field trips make for fun and memorable birthday presents. You can do the same for someone in your family, although you may prefer not to look for rattlesnakes.

Catching a Coachwhip in One Uneasy Lesson

Obviously, I study and enjoy working with reptiles. So to me, catching snakes is routine. You see a snake; you catch it in the appropriate manner; what else is there to say? Nevertheless, if you associate with snakes for very long, you will see mistakes made and make some yourself. Mistakes can be embarrassing. But if you get over the embarrassment, today's mistake can be tomorrow's anecdote.

In herpetology, as in other professions, you must learn the tricks of the trade to be really proficient. One trick that herpetologists learn, at least those who manage to survive, is how to catch snakes without getting bitten. You can see how this would be an important skill to have. Learning, however, is sometimes accomplished through trial and error. As one of my former students discovered, being told how to accomplish something is not a guarantee of success. I had taught Morton (not his real name) one of the tricks of the trade—or so he thought. The actual learning process came later.

Herpetologists have many different ways to catch snakes—ways that ensure they will never get hurt. Well, hardly ever. One method, called the sling technique, is excellent for capturing certain types of large, nonvenomous snakes. You grab the snake by the tail, sling it through your legs, and clamp your knees shut. Then, without a moment's hesitation, you pull it back between your legs, real fast. The instant the snake's head comes out, you grab its neck, right behind the head, with your other hand. No problem—a simple recipe for how to catch a great big nonvenomous snake of those species that show no hesitation about biting. Professional herpetologists can do the sling technique with a big black snake or water snake and not get bitten or harmed in any way. You have to be fast, but when carried out properly and smoothly the whole process takes one second. In fact, if it takes more than a second, you may have a problem. You cannot let a snake stay back there too long after you sling it through your legs, because some snakes tend to bite anything that's handy.

I show the sling technique to students in my herpetology class. But a definite distinction lies between seeing and doing. Most students never have occasion to try the technique, unless they become herpetologists. Following his classroom experience as a student in my herpetology class, Morton took a job as a wildlife biologist at a regional nature reserve. He had been shown the sling technique in class but had never done it himself. And my, what a difference between seeing and doing!

One spring day he was showing a group of schoolteachers through the nature reserve. Suddenly a seven-foot-long coachwhip snake came zipping across the path in front of them. Everybody squealed and ran for cover—except for Morton, who was leading the group. Coachwhips look like a thick piece of rope or a bullwhip. Unlike most snakes, they seem mean, biting at the slightest provocation. They are not venomous and cannot really harm you, but they will bite, with a big mouth full of small but pointed teeth. Morton apparently felt he had to do something to show that he was not like the screaming bunch of schoolteachers accompanying him. He knew the snake was a coachwhip and not venomous, so he decided to catch it. He thought he had learned how in the herpetology class. As the snake went by, Morton grabbed the last six inches of its tail and slung the rest of the body through his legs.

As you might imagine, a coachwhip would consider this to be more than slight provocation. Nonetheless, with proper execution of the sling technique there should be no problem. But there can be one. When you are holding a seven-foot snake by the tail and slinging the rest of its body through your legs, you have about six and a half feet of snake back there behind you. As you might also imagine, any number of things can happen—or get bitten. After he slung the snake through his legs, Morton made the mistake of waiting just a half-second too long. When he tried to yank it back through his legs, the snake had already come up behind Morton's head and grabbed him on the ear. And coachwhips do not let go; they chew. While Morton pulled on one end of the snake, the other end was hanging onto his ear.

That day provided several learning experiences. The schoolteachers learned a few words they did not know were in an ecologist's vocabulary. Morton learned a little about the sling technique and about the difference between knowledge gained through observing and that gained from experience. And he learned that wildlife biologists and ecologists, like everyone else, should remember not to take themselves too seriously.

In my own case, alligators have been extremely helpful in instructing me not to take my own position in the universe too seriously. They have proved to be experts at restructuring my thinking about who is really in charge.

Allez, Gator

One summer I was visited at SREL in South Carolina by Professor Donald W. Tinkle, a famous herpetologist at the University of Michigan. Four graduate students who had come with him had never seen an alligator in the wild. I assumed the role of the wise and experienced ecologist and confidently told these midwestern students that we would go out that night and catch an alligator. We went to a reservoir on the SRS and caught a small one with no trouble. The little gator did whap one graduate student in the face with its tail and bite another one on the finger; nothing serious. I smiled knowingly at their inexperience as we put the little creature in a sack.

Naturally, they wanted to catch a bigger one, and sure enough we were able to find a six-footer. I will omit a few lines on how we noosed a six-foot alligator and managed to keep its mouth shut and its tail from knocking everybody into the water; that procedure is routine and unexceptional for a herpetologist. By the time I crawled back into the boat, the students and visiting professor had used some nylon rope to tie the gator's mouth shut and its legs up over its back. It, like the smaller one, was now a helpless captive.

We returned to the dock and got in the van—three in the front seat, three in the back—with the two securely held alligators in the compartment behind the back seat. As we drove away from the reservoir, I was as warm and smug as could be expected of someone soaking wet. I felt the satisfaction of accomplishment, the feeling of being in control. Soon we were zipping along the deserted highway.

All of a sudden, one of the graduate students in the back seat said, none too calmly, "Hey, the alligator's loose . . . the big one!" No one wants to hear that. Not even a herpetologist. The student provided additional information: the rope had slipped off the gator's mouth. Furthermore, the six-foot gator was coming over the backseat.

Alligators are fast, but midwestern graduate students are faster. Before the alligator had cleared the top of the backseat, the graduate students had made new seating arrangements in the front. About the time the six of us had settled into this new format, barreling fifty miles an hour down the highway, another student (I believe it was the one on my shoulders) made a new observation at the top of his lungs. The alligator was now going under the front seat!

I guess that was better than hearing it was coming "over" the front seat.

The three of us in the front seat, with the students on our backs, did what anyone would do upon receiving this last bit of information. We lifted our feet. When I lifted mine, I accidentally put my foot on the light switch. We were now going fifty miles an hour down a dark highway with no headlights and with an alligator under our feet. I made a frantic but futile search for the light switch while we coasted in the dark for the length of a held breath. We eventually came to a prolonged halting stop, which is what would be expected in third gear in a straight-shift vehicle going from fifty to zero miles an hour. An added level of excitement is gained when this is done in total darkness and the clutch and brake are resting just above a six-foot alligator's head.

Someone managed to find his flashlight, and we looked to see what our fate would be. Fortunately, the alligator's head was too big to get all the way under the front seat and was lodged under the driver's side. The creature's nose was sitting about where my left foot would have been; its tail was draped over the backseat. We managed to dislodge the escapee, and I had three students sit on him in the back compartment for the rest of the trip back to the lab. Recovering our dignity was a different matter. It was fairly clear who had won this round and, once again, ecologists received further training in not taking themselves too seriously.

Skunks 6; People 0

Firsthand experiences can greatly increase our knowledge, and fortunately for the natural world, other species besides snakes and alligators do win some of the rounds against people. Consider my first encounter with a skunk.

During my college years, I worked one summer as a cook at a resort in Colorado. I was majoring in biology and felt obligated to profess a high degree of knowledge about the animal world—a degree of knowledge that I did not always possess. Sometime in the early part of the summer, I told the people with whom I worked that I knew how to catch a skunk. I had, indeed, been told about the skunk-catching technique by a biologist: all you have to do is lift up a skunk by the tail. Just lift it off the ground, and it cannot spray you. That is what I had been told. Well, I didn't really know if this were true, but it seemed a safe enough story inasmuch as my friends didn't know either. I did not mention that I had never even seen a live skunk.

One evening someone said he had seen some skunks at the garbage dump and suggested that we go catch one. It was kind of hard to bow out at that point; I had to appear eager. Six of us trooped down to the dump, and much to my chagrin one of the skunks was still there. I was put in the position of having to prove myself. "Show us how to catch it," my friends cried. Their enthusiasm was sickening.

An interesting feature of skunks is that they do not run very fast; they are pretty sure of themselves and not in much of a hurry. A person can walk almost as fast as they run. So as the lone skunk waddled away through the woods, with its tail up in the air, my friends and I jogged along behind it. I didn't really know what to do but finally decided to run real fast by the skunk, grab it by the tail, and lift it off the ground. Sure enough, I was suddenly standing there with this full-grown skunk. And it did not smell!

It did not do anything except thrash away with four black legs. So I thought, hey, this is really good. I was sort of a hero. Then it occurred to me that the biologist had not told me what to do next. After you have caught the skunk and are holding it by the tail, what do you do? Drop it off a bridge?

I remembered that outside the restaurant where we all worked was an empty fifty-five-gallon drum. I figured we could just open the barrel, throw the skunk inside, and put the top on. Again acting like an expert, I informed my friends that skunks sometimes make good pets. Maybe we could tame this one. So we headed back to the restaurant. Everyone gathered around the drum; we removed the lid; I said, "One, two, three!" and dropped the skunk right into that barrel. The only problem was, when I let go of the skunk, it spread out all four legs and with each of its little feet grabbed a side of the drum.

The last thing I remember seeing, before I was practically blinded, was this bushy black-and-white tail curled up even tighter than before and a skunk spinning around on that drum, spraying every single one of us. No one was spared. Two people got sick, no one could see, and I could hardly breathe. It was the most powerful, pungent smell I have ever experienced in my life, before or since. When I finally pried my eyes open, with tears running down my cheeks, I saw the skunk daintily climb down the side of the drum and walk back to the dump. It was pretty clear who was in charge, and it was not any of the six hapless people.

Fashion and Ecology Don't Always Mix

I do not recommend that anyone else try my skunk-catching technique without professional training, which presumably should not be learned from me. But I do recommend that you consider what you wear if you plan to have an ecological adventure. Ecologists, like people in any profession, have clothing styles and fashions that some take seriously. Many ecologists have distinctive hats (usually old); others have special shirts (also usually old, with memorable stains, strange drawings, or clever sayings that endear or inflame, depending on the audience). Some have a pair of lucky boots or a favorite belt buckle. At the time I was rather fond of an old, faded-and-patched pair of blue jeans I wore on field trips.

One reason for taking an ecology field trip, aside from the prospect of excitement and adventure, is to learn about nature, even human nature. On one particular trip I learned that Hollywood actor Tom Cruise apparently does not go on many ecology field trips. I acquired this intriguing insight after having observed Tom Cruise on two or three interview shows wearing blue jeans with a big hole in the knee. Now, we all know that a Hollywood movie star can afford new blue jeans and that the hole in the knee is a fashion statement. I do not know whether blue jeans are actually sold with holes in the knees, but I am observant enough to notice that a lot of younger people wear them in such a state. I can confirm that not many *real* field ecologists wear such attire on a regular basis.

One day I went with Vinny Burke, who was then a graduate student at the University of Georgia, on an unplanned trip to a field site. We were studying turtles in an aquatic habitat as part of our wetlands ecology research, and Vinny wanted to show me an area where he thought some of the turtles congregated. When we left for the field, I grabbed a pair of blue jeans, field shoes, and a grubby T-shirt from a closet. I hurriedly began changing out of my office clothes and putting on the field clothes as we rode along. The process went even faster after Vinny suggested that he drive.

Field ecologists depend heavily on blue jeans because they offer protection from a lot of biting, stinging, and thorny things. As I put on this old and worn pair, I ripped a huge hole in the right knee, just below an earlier patch. How fashionable, I thought. I will look good in the field, like Tom Cruise. I asked Vinny if he had brought a camera. I may have looked good (although possibly not as good as Tom Cruise), but I would soon feel awful. I first realized that fashion and ecology do not necessarily go hand in hand as we made our way from the truck to a pond. We walked through an enormous patch of blackberry bushes about the size of Vermont. Blackberries are the silver lining of a very black cloud known as blackberry briars. By the time I reached the water, my right knee looked as though an angry bobcat had used it as a scratching post.

Well, I thought, once I start wading around in the water my knee won't feel so bad. This was momentarily true until we reached an area of knee-deep water. Then I discovered a second hazard of a good-looking tear in blue jean knees: a small insect known as a backswimmer. Backswimmers live in many aquatic habitats of North America, and they defend themselves by jabbing their mouthparts into the body of anything they consider a threat. They are not venomous and will not hurt you at all if you do not provoke them in some way. But when they "bite," it hurts as much as a bee sting for a minute or so.

Anyone wearing blue jeans, socks, and shoes is generally immune from bites by backswimmers, because they only bite people when they are pressed against some part of the person's body and cannot escape. Do you think a big hole in a pair of blue jeans would let in a half-inch long backswimmer? Yes, quite a few of them. One after another ended up inside my pants leg, pressed tightly against my right leg. Being squeezed between denim and tender flesh provoked them, and each proceeded to protect itself. The experience was not enjoyable. (Tom Cruise always seems quite content in knee-torn blue jeans when he is being interviewed. But, of course, I have never seen him interviewed while he was standing in water infested with backswimmers.)

After we left the water, cleverly avoiding the blackberry patch, I encountered a low-lying broken branch that aimed right for my Achilles' knee. Later, a horsefly decided my blood-stained, exposed knee was the most delectable eating spot in the county. The final blow came when I knelt down to observe a hole in the sand dug by a racerunner lizard. Sure enough, the only sharp-edged rock within a hundred yards was lying right where I put my knee.

I plan to discard this fashionable pair of blue jeans. If Tom Cruise needs another pair, he is welcome to them.

7

In Your Own Backyard: Exploring Nature's Wonders

A large insect attacks a power tool. Wasps dance in response to human movement. One bird calls another to come see a snake eating a mammal. Sound like a cross between a Disney film and science fiction? Actually the events were just routine happenings that could occur in anyone's backyard. So happens, they occurred in mine.

The insect attack came one evening as I used a power saw to cut a board—a routine householder's chore, until the saw blade hit a knot in the board and emitted a high-pitched buzz. Suddenly, what I took to be an enormous horsefly hit my shoulder, whirled twice around my head, bounced off the handle of the saw, and disappeared into the backyard vegetation. I hit another knot: another zinging sound from the saw blade, another attack. This time the attacker landed on my arm long enough for me to see that it was a cicada. At that moment, it occurred to me that the sound of the saw mimicked a calling male cicada.

Male cicadas call to attract females. I was presumably being courted by a female cicada that thought she had located the stud muffin of the cicada world. Or could this have been another male responding to what he thought was a superstar competitor? I don't know if male cicadas behave that way (nor did the five entomologists I called at three different universities), but some male frogs act aggressively toward other calling males. Whatever the case, I brought family members out on the porch to watch the performance. Saw a knot; attract a cicada. The show was impressive, predictable, and worth the price of admission.

The dancing wasps occasioned a minor family disagreement. I was trimming the bushes at the front of the house, a task my wife, Carol, supported more than I did, but the real controversy came with my discovery: a small paper wasp nest in one of the bushes. Carol wanted it and the wasps removed, and she soon held a can that could grant her wish. I interceded, pointing out that having a wasp nest eighteen inches from the porch steps might reveal some interesting insights. Besides, they had not stung me while I trimmed away the leaves above their home. Carol reluctantly agreed not to exterminate the wasps unless someone got stung (secretly hoping it would be me, I'm sure).

The next morning, while the day was still cool after an evening rain, I was vindicated. As we watched the nest, Carol discovered that when she moved her head a few inches, the five wasps standing on the nest would reorient themselves. If she moved to the left, the copper-colored wasps would sidestep in unison to face her. When she moved back to the right, they would shift back to their original position. They were perfectly synchronized as they stood on the nest, always facing directly toward her.

And what of the aforementioned birds, snake, and mammal? Whenever you hear the raucous cries of blue jays, investigate. You have a good chance of seeing some interesting form of wildlife. One summer day, upon hearing a blue jay summons, Carol found a small corn snake crawling along the backyard path. The birds were not happy. After catching the snake, we returned it to the backyard that night, when no birds were looking, except maybe one—a screech owl whose presence had been announced by the jays earlier that week.

Another backyard event was more dramatic. Hearing the jays, we stood on the porch to see where their attention was focused. Below a bush, we saw two wrens and a brown thrasher fussing at something in the ground cover. Amid the monkey grass lay a gray rat snake with a small rat in its coils. We backed off to let the drama unfold. Ultimately the snake swallowed the rat and crawled out on the limb of a small tree. The blue jays eventually tired of the sport, but the wrens fidgeted around the area for another hour. The snake, unable to hear all the commotion in its behalf, had seemingly ignored all the bird excitement and had gone about its business.

The natural world is action-packed. Nonstop, real-life dramas unfold constantly, all around us—including our own backyards. All we have to do is look, listen, and learn.

Why Not in Your Backyard?

Today's technology allows us to probe into biological mysteries not even hinted at two or three decades ago. Yet basic questions about some of our most common animals still go unanswered. We still have a lot to learn. In one sense, our ecological ignorance should come as no surprise. Paul D. Haemig of the University of Umeå (Sweden), writing in the *Bulletin of the Ecological Society of America,* declared that Americans are "unable to identify the plants and animals in their own backyard." As pointed out in the book *Keeping All the Pieces,* he may have been overstating the case somewhat. After all, few if any biologists anywhere, especially commentators on TV nature shows who delve into the intimate lives of exotic animals, could identify every one of the hundreds of small plants and insects that dwell in a big yard. But the assertion rings true in some respects. An appreciation of natural history has seldom (if ever) been foremost among most residents of the United States.

The deeper meaning in Haemig's message is that many Americans do not seem to care enough about their own local environment to find out what lives around them. Not caring what creatures inhabit your yard can give the impression that you have no concern for their welfare. Perhaps you have no such concern, but if you do, an opportunity exists to demonstrate it.

The Backyard Wildlife Habitat program sponsored by the National Wildlife Federation (NWF) deals with the issue in a direct, hands-on (for the participant) fashion. Operating on the principle that more natural habitat means more native wildlife, the NWF is a leader in teaching people how to enhance wildlife by improving habitat. The program helps people plan a wildlife habitat in their own yards. Once certain straightforward guidelines are completed, official certification can be obtained by completing an application form and sending a fifteen-dollar fee. A small investment for a substantial return.

The program emphasizes four points for developing and maintaining a backyard habitat to which wildlife will be attracted: the basic ecological requirements of food, water, cover, and places to raise young. The widest variety of habitat elements are recommended to attract the greatest diversity of birds, insects, and other animals. The program also emphasizes the value of using native plant species as much as possible to provide food sources. Plants that bear acorns, nuts, or berries are obvious choices for providing food for many species. A nature center or nursery can recommend suitable trees and shrubs for an area.

The most difficult essential for many people to provide is water. The program suggests birdbaths, a dripping hose, or even a small sunken pool kept as a year-round water source. Supplying running water could be costly, so one would have to decide if the returns on wildlife attracted and sustained justify the expense. In many cases, the answer would be yes.

Cover for wildlife should be the easiest item to provide and can even be a blessing to anyone tired of carting away shrub and lawn trimmings. Piles of brush or rocks need not be discarded; they can be organized aesthetically, in a manner that will satisfy neighbors and attract wildlife. Creating hiding places for small animals will greatly increase biodiversity in your own backyard.

The final ingredient for a healthy and diverse backyard is the creation of places for animals to give birth and raise their young. Dense shrubbery, tall trees, and some of the wildlife cover will help in this effort. Although some species are extremely particular about where they nest, placing birdhouses on trees will be sure to attract something. Leaving a dead tree with holes in it can provide nesting sites for woodpeckers, flying squirrels, and lizards. (If the tree poses no danger from falling, why pay someone to take the tree away?)

One recommendation in the Backyard Wildlife Habitat bulletin is titled "Cut Your Lawn—In Half." Lawns offer little of value to wildlife, and they require great amounts of water to sustain. The bulletin suggests reducing lawn size by planting shrubbery, trees, and flower beds. Another heading reads "Let Predation Reign—Pitch Your Pesticides." The point is made that "in the natural world, virtually everything is sought out and 'eaten' by something else." Advice is given for ways to reduce the use of insecticides, fungicides, and herbicides.

To find out more, including how to have a yard certified, write: Backyard Wildlife Habitat Program, National Wildlife Federation, 1400 16th Street N.W., Washington, D.C. 20036-2266.

Bringing Moonbeams Home in a Jar

Whether or not you have a backyard habitat that qualifies for certification, any time is appropriate to be reintroduced (or introduced) to nature, to arouse an appreciation of the natural wonders that surround us. Start in your backyard, a park, or a wooded area in the neighborhood. If you get up early enough and have trees and shrubs near your home, you might hear the dawn chorus. You do not have to name the birds to enjoy their varied songs, just as you can appreciate a choir without knowing anyone's name. Morning-time birds seem to gossip, chatter, and scold. Attend to the birds and wonder what they are saying with their complex calls.

Or maybe you want to start after dark. Take a flashlight, but do not use it unnecessarily. The woods and fields in many areas are alive on summer nights with animals that carry their own lights—lightning bugs. Anyone not intrigued by these sparks of the summer forest needs some time off. Bringing moonbeams home in a jar is easier than it sounds when the fireflies are out and you have an empty jelly glass. We watched them in the backyard one night, right before a rainstorm. The woods twinkled. As the rain came, the yellow lights disappeared, leaving a dark woods suffused by falling rain with a beauty of its own. Later, the rain ceased and the woods were once again alive with flickering points of light. We noticed the lights were all high in the trees, not low as before the storm. Had they sheltered from the rain beneath the largest limbs in the treetop canopy? Minutes later the flying males were nearer the ground, continuing on their mission of looking for the female glowworms that lay among the now wet leaves.

The excitement of nature is in the soil and waters as well as in the air. Turn over any log or flat rock and if the season is right you will likely find something of interest. Taking someone with you—particularly a child—can enhance the wonder and delight experienced on a nature expedition. Some people have been known to enjoy the sights and sounds of nature even more than television or video games. You can also touch, smell, and taste nature, and that is better than virtual reality.

On an excursion with two neighborhood children, we turned over a board for our young friends, Andrew and Emily, who were pleased with our finds: three slate-gray roly-polies, a white beetle grub, and a little orange, not-to-be-touched, centipede. Children should learn not to touch any animal unless someone who knows says it is safe to do so. They should be taught not to fear nature, but to respect it. Encourage a child to pick up a roly-poly or a white grub but not a centipede. Centipedes like to be left alone; so do grubs and roly-polies, of course, but they do not have biting pincers with venom in them.

No animal, regardless of whether it can bite or sting, wants us to intrude on its life, so children (and adults) should learn to return animals they pick up. Emily wanted to keep the fat, white grub, which is the larval stage of a type of beetle known as the bessie bug. Bessie bugs, which are long, black beetles found in rotting wood, are ideal pets for children—a harmless insect to watch and handle. But Emily put the grub down after Andrew returned the roly-polies to their home beneath the board.

What about keeping an animal for a pet? A fine idea, in

some instances, if certain guidelines are applied. The first rule to remember is that these days many of our wildlife species are protected in a manner that makes it illegal to capture them at all. For example, species officially declared endangered or threatened cannot be harassed or harmed in any way. Almost all birds in the United States are illegal to catch, and the laws in some states are very strict about removing certain amphibians, reptiles, or mammals from the wild or keeping them as pets. The animal must be brought home only as a visitor, not as a permanent resident. No venomous creatures should be acquired as pets for children (or for adults unless you really know what you are doing). And the visiting creature should not be something that can easily escape or that requires constant feeding. Another rule that most parents would sanction is to set a visitation limit for the new pet. After one day, a week, or some other agreed-upon time limit, the animal is returned to its own home.

Spring, summer, even autumn are times to enjoy flowers. Take a walk and savor the fragrance of flowers, shrubs, and trees. Ponder the mystery of why their perfumes are different. Examine each flower to see the many kinds of insects flitting around. A flowering shrub or tree can harbor a wonder world of insects.

Visit a lake, river, or stream. The shore life is probably abundant, although perhaps not apparent. Water birds, turtles, and frogs have obvious appeal, but smaller animals and plants can also be engaging: water bugs or flying insects, lily pads or algae. Anything can be interesting if examined closely, with a questioning eye.

One way to appreciate the diversity of plant and animal species that inhabit a region is to speculate about the lifestyle of each one. How does an earthworm get its food? How does a slow-flying monarch butterfly avoid being eaten by birds? Why does a male box turtle have red eyes? The answers to some questions may be found in books. Sometimes, however, no one knows for sure, so your guess may be as good as any. Question why a jay is blue and a cardinal red, why a skunk smells bad but honeysuckle smells sweet, or why a few snakes bite but most never do.

Much of our wildlife and natural habitats in North America are still intact. Many people place a higher value on these natural resources than on enterprises that degrade or destroy them. Was Robert Louis Stevenson referring to the world's wealth of biodiversity in his two-line poem "Happy Thought"?

The world is so full of a number of things,
I'm sure we should all be as happy as kings.

A Class Act

To ensure a legacy of high biodiversity we must be certain the younger generation learns to appreciate and value all species, although sometimes one might wonder if adults should be the ones in class learning about ecology. The children seem to have learned a lot already. This point was admirably demonstrated at the touchstone of our nation's environmental attitude, a classroom of fifth-graders. We spent two hours discussing ecology with the students. Do fifth-graders seem like unlikely candidates for a touchstone? Maybe to some, but today's fifth-graders will be the college graduates and leaders of society in the early part of the twenty-first century. What they think about issues could soon affect our whole nation, indeed our entire world.

The class assignment that day was for students to report

on the weekly update of their environmental journals, which consisted of newspaper or magazine articles they had read about environmental issues. Each student had selected a current article about the environment and had presented a summary to the class. We were pleased with the students' views on ecology; they revealed a high level of environmental awareness and education. The diversity of articles was also impressive. Of twenty-six reports, only two students reported on the same topic. Twenty-five different articles about the environment—clearly, the news media find the environment more worthy of coverage than they once did. Topics included a program in Indianapolis to plant a tree for each baby born, an administrative appointment at the Environmental Protection Agency, and tips on how to make recycling trash more fashionable.

Every student in the class gave a report, which indicated to us that the students considered compiling an environmental journal a worthwhile exercise. How many other homework assignments have a 100 percent completion rate? We discussed each topic presented. One was that bald eagle nestlings in a protected nest on the Department of Energy's SRS in South Carolina had survived a winter storm. Everyone knew that at the time bald eagles were an endangered species. One student emphasized that even to take a feather of an endangered bird is illegal. He said that in order to use real eagle feathers in costumes for the movie *Dances with Wolves,* the producer had to rely on tribal hand-me-downs acquired before passage of the Endangered Species Act.

Another report was about cereal boxes made from 90 percent recycled paper, of which 35 percent was postconsumer waste. Every child knew that "postconsumer waste" means paper products previously used and discarded and that it differs from another form of recycled material—"manufacturing waste," cuttings and trimmings from the factory floor. A higher proportion of postconsumer waste means higher efficiency in a recycling program. Obviously, cereal boxes need not be made from freshly cut trees. Today's young people would consider such a process totally unnecessary and unacceptable; for them, recycling has become a way of life, the only environmentally acceptable way of life.

One article was on drought problems leading to famine in parts of Africa, and we discussed two of the possible causes of famine. One was deforestation, which may magnify droughts in some regions, and the other was human overpopulation. We illustrated the problem of overpopulation by offering the scenario of the fifth grade going to lunch and discovering that the sixth grade had also shown up to eat at the same time. Simple formula—if the amount of food stays the same, each individual has less to eat.

Someone read an article about an amusement park development project that could result in the closing of a coastal shellfish industry. When asked if they would enjoy having a few more fun houses and water slides, a student replied—unprompted, it should be noted—"We have enough of those already." Remember, this is fifth grade.

The students' overall level of environmental sophistication and presentation of the issues was impressive. The performance of a class, of course, reflects in part the efforts and guidance of the teacher, but students also emulate their parents' views. Plus, everyone is influenced by the community in which they live. Children's attitudes are a gauge of the opinions, behavior, and teachings of those around them. We left the school with three strong, and very positive, feelings—the teacher was doing an excellent job of environmental education; the environment is now recognized as important by many people, as evidenced by the wealth of environmental information presented; and we can have confidence in today's youth as trustees of the natural environments we will leave them.

Anyone who teaches elementary school—or English, journalism, or science in the upper grades—might consider initiating an environmental journal project. It is an assignment that students can be depended on to complete.

Ecology Scavenger Hunts

Ecology assignments need not be restricted to school. For a game at home, find one beetle and a green leaf with points on the edges. Add something that lives in the water, a bird with red on its head, and a creature with more than six legs. Sound like a recipe for the bubbling cauldron in *Macbeth*? Well, it is a recipe, of sorts, but not for witches. It's a recipe for making people more aware of their environment by spending more time outdoors. We call it an "ecology scavenger hunt." Originally, we thought of it as an activity for children, but our cousin Anita and her husband, Wayne, disabused us of that notion. Staying in the mountains near Gatlinburg, Tennessee, after Christmas, Anita, Wayne, and another couple went on an ecology scavenger hunt in the woods outside their chalet. They never found a water creature or a bird with red on its head, but that shortfall in no way diminished their pleasure in the hunt.

One reason we think an ecology scavenger hunt is a good idea is that many people spend too much time indoors—television and computers make it easy to do. Indeed, these are important advances in technology; a person can learn a lot from TV and CD-ROMs. But the living world is outside, and people should be encouraged to spend as much time there as they can. The idea of the scavenger hunt is to catch or see each item on the list. This means spending time in the yard, a park, or the nearest woods. An additional requirement of the hunt could be to read something about the plant or animal, even if it means a trip to the library, museum, or bookstore.

Anyone should be able to find a beetle, even in winter if the ground is not covered with snow, and find something written about these animals. What kind of beetle is it? More species of beetles live on earth than any other major group of animals, so it is easy to find something to read about beetles. (Which reminds us of the best beetle quote we know: J. B. S. Haldane, the outstanding British geneticist, was being challenged about his religious beliefs. When asked what he believed about God, Haldane replied something to the effect, "I believe He had a certain fondness for beetles.")

As for the other ecology scavenger hunt items on the list, what kind of tree or shrub did the pointy-edged leaf come from? The reading might be about evergreen trees. Any body of water will have plenty of living things to see. Some of the algae and zooplankton may be invisible without a microscope, but with a careful survey of the edge of any pond or stream you will find something alive. You might be amazed at just how many living organisms are there, even if they are not all moving. As for the bird on the list (which might at first glance seem to be the hardest item to find), one might have to go to a book to learn which birds qualify. Maybe you have a red-headed woodpecker in the neighborhood. Lots of other woodpeckers have red on their heads (as do ruby-crowned kinglets and turkey vultures), but the color is sometimes hard to see. Learn to identify these birds from a bird field guide. (If you think about this challenge, however, you will realize that "red on its head" does not exclude having red all over its body. Almost everyone can identify a cardinal.)

And for the final item on the list, what has more than six legs? Millipedes, centipedes (which you should *not* pick up because of their venomous bite), and spiders, to name a few. All of them are present in the woods, maybe even in the backyard, under rocks or logs or tree bark.

This list for an ecology scavenger hunt may be a bit difficult for where you live, where you vacation, or where you have a family reunion. If so, take a stroll around and make up your own list. An ecology scavenger hunt is a good excuse for being outdoors—a place with which everyone ought to become familiar. In fact, getting ourselves outdoors and more familiar with life around us provides one of the two most important ingredients for saving endangered species and preserving natural habitats: awareness. The other essential is money, but without a donor's environmental awareness, money may be difficult to obtain.

Another Class Act

Cultivating the idea that we must save our wildlife for the next generation is critical for a healthy environmental future. Such an outlook is imperative in one especially important group of people: children. After all, their generation's future is at stake. We learned of an excellent approach for creating environmental awareness of endangered species among schoolchildren. It is an approach that could be adopted by members of adult organizations as well.

The students of a school vote to "adopt" an endangered species found in their state and pledge to help with its management and protection through a fund-raising campaign. This strategy has been used by students at Long Junior High School in Cheraw, South Carolina, whose choice was the then federally endangered bald eagle. Collection jars were set up in homerooms where students could unload some of their spare change. For a six-week period the principal announced over the school intercom the total amount collected each day. The program was encouraged by the Nongame and Heritage Trust Section of the South Carolina Wildlife and Marine Resources Department. Money collected by the students directly assisted the endangered species programs in the state. In the case of the bald eagle, the funds helped refurbish a flight facility used to rehabilitate eagles, hawks, and owls.

Programs such as this one develop in students (as well as teachers and parents) a broad understanding of wildlife and habitat issues. For example, to select an endangered species, students must be aware of the status of different ones. Some may be protected at the federal level, whereas others may only be recognized as endangered by the state.

For schools or other organizations interested in developing such programs, the first step is to identify the species that are candidates for assistance. To find out which species are protected by a particular state, obtain a list from the state's fish and wildlife or game department. The exact name of the agency varies by state, but it is almost certainly located at the state capital. We obtained a list of Florida's protected species from the Florida Game and Fresh Water Fish Commission's Division of Wildlife in Tallahassee, which indicated a willingness to recognize a program like that set up by Long Junior High School and acknowledged that the donated monies would be put to good use. We feel certain the wildlife division of any state would be equally cooperative.

To find out what species in a region are protected at the federal level, you can obtain a complete listing of all endangered and threatened species by writing to: U.S. Fish and

Wildlife Service, Publications Unit, Room 130 Webb Building, Arlington, Virginia 22203, or by calling (304) 876-7203.

An endangered-wildlife program at a school could develop into some fascinating projects. Imagine a campaign with posters, speeches, and letters to the school newspaper as students justify why a particular species should be chosen. For example, students in Alabama or Tennessee might champion the pale lilliput pearly mussel: "When you help a mussel, you help the whole river." Students in New Hampshire might choose to promote protection of the federally listed puritan tiger beetle: "They may be tiny, but they need our protection, too." Or in Minnesota they might say, "Sound off for the piping plover. Protect our birds."

Such a program could have applications across the curriculum. Science classes are an obvious forum, but projects could also involve journalism, art, and library research. Term papers for English class could be based on the biological background, geographical range, and historical record of particular species.

Beverly King, the librarian at Long Junior High School, neatly summed up the importance of programs such as these: "Our students want to show that they are concerned about the plight of endangered species, . . . but even more than that, they want to do something about it." They *are* doing something about it, by collecting funds and by creating a proper awareness of the importance of the natural environment.

Vertical Ecosystems

Sometimes generating a proper awareness just requires looking at the ordinary, the mundane, the everyday from a different perspective. One of the most common life-supporting systems in the world is an ecosystem found in downtown London, midtown Manhattan, within national parks, along many oceanfronts, and in your own neighborhood. The ecosystem is one we see daily but seldom think of in ecological terms. What is it? The answer: Walls. Yes, walls. Like those around a garden or the sides of houses and sheds. Arnold Darlington, in a book called *The Ecology of Walls,* claims that walls comprise more than 10 percent of the area habitable by plants and animals in a city.

Many factors affect the extent and composition of species inhabiting walls, including the degree of inclination. Walls at a more horizontal angle, or with shelf space, are more likely to collect dirt and debris where seeds can root. The compass direction the wall faces could matter for some species. Would moss be more likely to grow on the north side of a wall? The material, porosity, and composition of the wall, the climate of the region, and the history of human alteration are also major influences on what is found living on a particular wall.

One influential factor determining the vegetative character of walls is age. Darlington says the best mural vegetation, as he calls it, is found on walls more than 150 years old. Our Ivy League schools come to mind as a good example. When walls get several centuries old and are left unattended, as with two-thousand-year-old walls built by the Romans in many parts of Europe, they become badly decomposed. Then one is most likely to find shrubs and trees growing from the wall ruins, a condition equivalent in some respects to a climax forest, a plant community that has successfully adjusted to its environment. In a study of walls in England, researchers found that algae and lichens were usually the first pioneers to become established. Then various mosses and sometimes ferns took root, followed by other kinds of vascular plants. Vines of course can root at the base and climb. Other seed plants must find a foothold in collected dirt or in the rotting organic matter from a previous inhabitant. Once a wall has structure in the form of vines or other plants, or has developed crevices, animals begin to take up residence.

The ecological perspective of walls offers some new and intriguing prospects. School projects come to mind, or just a form of entertainment by looking at the world from a different angle. We decided to take a look at some walls around our own yard. An old concrete block incinerator had algae and lichens at the base near the ground. We didn't find any moss, on any side, but did find signs that two animals had been there. One indicator was a series of hollow mud structures made by females of the black wasps called dirt daubers. The mud nests contained spiders that had been captured and paralyzed by the dirt dauber before she laid her eggs. (The first meal for a young dirt dauber wasp is a spider.) The other animal sign was a spider nest covering a depression in one of the concrete blocks. The spider was missing—a victim of dirt daubers?

Along the brick wall of the house, we found not only the obvious ivy and Virginia creeper but also a variety of small creatures, including some animals on which we could pin a name—aphids, a millipede, and a caterpillar—and a few insects we could not identify. Imagine the simple ecological principles and processes that could be examined as part of a science fair or classroom project. Do wood, brick, and concrete walls in an area differ in the number and kinds of plant and animal inhabitants? Does a shaded wall have more organisms than a sunny wall? How important are the wall's age, height, or position relative to ground vegetation in determining what grows on the wall?

To sample a little wall ecology of your own, see how many different kinds of plants and animals you can find on walls in your neighborhood. Upon reflection, we realized some of our previous observations of lizards and snakes crawling, bats and tree frogs sleeping, and birds building nests all had one thing in common: their activities occurred on some sort of wall. The thought came to mind, why are so many plants and animals of flat regions already suited for living in a vertical ecosystem created by human beings? We took a look at the trees and realized that the forests are full of walls. Plants and animals have been using them much longer than we have.

The Urban Naturalist

One can step even further into the realm of human-made ecosystems. What do skyscrapers and nighthawks, city parks and reptiles, freeways and mulberries have in common? The answer: Each other, within an urban environment. Most general-audience books on ecology emphasize natural habitats and native wildlife. *The Urban Naturalist* by Steven D. Garber focuses instead on the plants and animals that inhabit cities and suburbs. Garber addresses which species are present, how they got there, and why so many are successful. This book does not concentrate on the negative aspects of an overpopulated world whose natural habitats diminish daily. Instead, it is "designed to help everyone view the urban landscape and its inhabitants as part of a vital, busy, intense, yet natural habitat." We applaud the author's purpose, although we disagree about whether downtown Manhattan qualifies as a natural habitat.

Garber's book is organized into chapters on major groups of organisms with which everyone is familiar, including grasses and wildflowers, trees, insects, fish, amphibians, reptiles, birds, and mammals. Each chapter discusses the ecology of several species, including exotics, found in urban environments. An example is the white mulberry tree, a species formerly native to China and a prime food source for silkworm caterpillars. According to Garber's book, King James I had mulberry trees and silkworm eggs sent to the Virginia colony in 1623 with hopes of developing a silk industry. As a consequence of the planting of thousands of white mulberry trees, silk production was profitable in America for a few years in the mid-1600s and again in the 1820s. By the twentieth century the silk industry had disappeared as a trade, but the white mulberry tree, "a decorative and urban-tolerant tree," had taken a foothold. Birds eat the mulberries; hence they spread the seeds. The species flourished in cities, perhaps due to reduced competition from native tree species.

Garber's point is that urban areas are full of trees and other organisms. The plants and animals that exist in cities are as successful as those in natural areas, sometimes even more so. People who live in urban habitats should learn to appreciate the flora and fauna around them in the same way that one appreciates fauna and flora in an undisturbed forest or wetland.

Not all urban species are exotics, and Garber emphasizes the persistence of many native species in cities. For example, painted turtles bask on logs and shorelines of lakes in many eastern and northern cities. Although not mentioned, slider turtles in southern bodies of water are seemingly oblivious to whether the water is in a city park or an unpopulated forest. Birds, of course, are common in cities. In suburban areas with abundant shrubs and trees, bird species diversity can be especially high during spring and fall migrations. Garber discusses a variety of native bird species that have adapted to urban situations: herons that feed along the edges of ponds, swifts that nest in chimneys, and nighthawks that feed on insects around city lights and nest on rooftops. Even wild mammals live in urban habitats. Gray squirrels are a featured species in many parks in the East and South. Opossums, in part because they can subsist on roadkills and garbage, are now found in most U.S. cities, including all parts of New York City. Raccoons, rabbits, and chipmunks can get by in suburban areas with the proper mix of woodland cover.

Many people view some urban species as less desirable than white mulberries, painted turtles, and gray squirrels. Norway rats, roaches, and pigeons come to mind, but Garber even has a point about some of these animals: "If we could get over our fear of rats, we might not have to pour so much poison down their holes and endanger so many other animals." In his view, "Rats are basically just another rodent, without a bushy tail."

Overall, *The Urban Naturalist* provides a positive environmental perspective for a city dweller: Find and enjoy the plant and animal life. Garber maintains that "the most rapidly growing habitat in the world" is the urban environment. Although we commend any of his readers who want to become urban naturalists, the ability to discover life in the cities does not justify complacency about the destruction of native wildlife and habitats that are decidedly more natural than Manhattan.

8

Time to Put Out the Night Light: In Our Opinion

We were driving along a lonely, peaceful highway. The sun had set and the night sky was a black blanket speckled with stars. The night's coolness drifted through the open window along with the sounds of tree frogs calling from nearby swamps. We enjoyed the serenity of a southern night while traveling beneath the starry sky. At a four-way stop in a small town, the spell was broken by an all too common sight—an empty shopping center ablaze with mercury lights on forty-foot poles. The stars disappeared, and we felt a sense of loss. Today's obsession with night lighting had stolen the natural beauty of a night sky.

One trait of human nature is to accept, often without questioning, phenomena to which we grow accustomed. Change may be so gradual that we do not notice or complain. Environments become polluted right before our eyes. By the time pollution is of a magnitude that warrants complaint, commercial forces may be difficult to combat. For some reason, a great part of our society seems to think that perpetual daylight would be a wonderful idea. Certainly artificial lighting has resulted in positive returns and progress. The practical benefits, allowing us to work or play after the sun has set, are obvious. But outdoor lighting can have negative ecological effects on plants, animals, and even people. While we have all watched, a problem has developed: photopollution. Today's society seems to be obsessed with having us light everything but cigarettes, almost as if we believed humans were meant to live in perpetual daylight.

Although some may be subtle and not easily recognized, the effects of outdoor lighting on plants and animals are extensive. For example, sea turtles are reluctant to nest on lighted beaches. If they do, the young suffer. When hatchling sea turtles leave their nests (normally at night), they become disoriented if artificial lights are near. High mortality from predation or dehydration can occur if the hatchlings do not reach the surf in a timely fashion. In one study, many traveled away from the water and some actually died in traffic on a busy street.

Migratory birds traveling at night can become disoriented in the vicinity of brightly lit areas. A study in Hawaii documented that large numbers of rare seabirds are attracted to lighted areas and can die as a result of this unnatural behavior. Countless insects perish nightly as a result of outdoor lights. Most of those affected, including beautiful moths and other bizarre and fascinating winged creatures, have no direct negative effect on humans. However, they play an important role in the environment as prey for other animals, as pollinators, or in ecosystem roles we have yet to explore. A nighttime of batting their heads and bodies against lights can't be good for bugs, even if it doesn't kill them.

Consider also the impact that synthetic lights have on human beings. Some practical problems concern potential detrimental effects of fluorescent lights and driver distraction at night. Anyone who has seen the Milky Way from the vantage point of an unlit desert or abandoned beach becomes aware of how city lights rob us. The work and research of many of the nation's large observatories are threatened by the night-sky illumination as cities increase in size: the lights that follow urban development make astronomical observations more difficult. To those interested in natural phenomena, the loss of the night sky is becoming too widespread.

We are all aware that we need lights in some situations, such as on dark city streets, to see potential enemies. But do most residential areas and after-hours parking lots really need so many lights that produce, from a block away, the illusion of an early sunrise? What is the appeal of a forty-foot mercury lamp in an otherwise serene countryside? Power companies and businesses specializing in outdoor lighting may take a more commercial viewpoint, but is a five-acre, empty shopping center with dozens of glaring lights at midnight really necessary? Excessive and unnecessary outdoor lighting is an environmental issue for which the costs and benefits should be weighed. The commercial viewpoint is understandable, but perhaps companies should seek ways to illuminate necessary areas in less obtrusive ways. One way is with the subdued, yellow "crime lights" used in some cities. These lights reduce glare and yet seem to function effectively in minimizing street crimes. Sometimes, placing lights closer to the ground can provide the necessary illumination without detracting from the surrounding habitat.

People in residential areas who put tall lights in their yards might use sensor lights that do not stay on constantly. We know of no evidence that a backyard light keeps thieves away—it merely shows them where the darkest shadows are. Having sensor lights that come on unexpectedly would be a far greater deterrent to an intruder than a predictable lighting scheme. Outdoor lighting may be something we have accepted without thinking about its effects or its true usefulness.

It is time to give the issue a little more thought, to stop being paranoid about the dark. Look around some night at the lights in your town, along your roadways, or even in your yard.

Then look up at the night sky and try to find the smaller stars you remember seeing as a child. Do you think we may have become a bit excessive in our efforts to light up the night? The natural beauty of the night sky is being stolen from us by the nocturnally challenged. We are not opposed to lighting up the darkness, but we are suggesting we keep it at one candle when that's all we really need. Where is an organization called DELITE (Decrease Excessive Lighting in the Environment) when you really need one?

Who Cares about Environmental Apathy?

A fight against unnecessary night lighting may seem trivial or even whimsical in comparison to other socioenvironmental concerns of today, from child abuse to terrorism, schoolchildren killing each other with handguns to violent criminals being released early from overcrowded prisons. Does destruction of the home of Myrtle's silverspot butterfly, an endangered species in California, matter when millions of dollars of real estate are being burned to the ground or buried by mud slides? Should the welfare of the ring pink mussel, a West Virginia river clam, really concern us when unemployment lines are longer than the river? Is it important that Florida panthers or Yellowstone grizzly bears live on when millions of human beings are dying from AIDS, cancer, and heart disease? With all these problems facing us, one might ask why we should be concerned about too much night lighting or any other part of the environment. These are legitimate questions that deserve an answer. In each case, the answer is yes.

A compassionate person can and does care about a variety of issues simultaneously. Caring about environmental issues need not diminish sympathy and support for humanity's social, economic, or medical challenges. Human and environmental concerns are not mutually exclusive. They are, in fact, inextricably entwined. An endangered species, whether butterfly, clam, or major carnivore, is symbolic of environmental deterioration. Each is an omen of environmental problems. Recognizing the environmental situation that has led to a species' impending extinction is of long-range importance to all humans beings.

California fires and mud slides cause major economic havoc and are of great concern, affecting thousands of people directly and indirectly. They may also be indicators of environmental deterioration. Overdevelopment, artificial suppression of natural fires, and alteration of vegetation patterns can all result in environmental instability. Concern for a butterfly's natural habitat does not minimize the seriousness of the loss of human homes. It should, instead, focus attention on what ought to be a universal concern: degeneration of the environment—the butterfly's and ours.

Unemployment is distressing and is always too high for those who want to work but have no job. A healthy environment could provide more jobs in the long run than one that has been despoiled. Hence concern for an endangered clam is not trivial; its endangerment signals an unhealthy river environment. Healthy rivers, forests, and oceans assure a higher quality of life for everyone, whatever their job situation.

Threats to human health are in some ways more forbidding than ever before. The risk of a debilitating or lethal disease is no longer something that happens only "to other people." Much of humankind's hope lies in medical advances, and many discoveries that promise cures and treatments for human diseases are by-products of plants and animals. Two examples: a fungus that grows on yew trees in old-growth forests of the

Northwest and the secretions from skin glands of a frog in tropical America. Yew trees and the fungus that grows on them produce taxol, a drug proven more effective in curing ovarian cancer in some women than any other known treatment. Secretions from the dart poison frog, used by Ecuadorian Indians to coat the tips of weapons, have been tested on mice and found to be "200 times as effective as morphine in blocking pain." We do not know now if panthers and grizzly bears have anything to contribute to human health care, but if they go extinct we will never know. Every species matters, and we should prevent the disappearance of as many as possible. A species we keep extant may eventually serve us far better than we have served it.

Perhaps the most important reason we should care about the environment is that natural habitats and wildlife are an essential foundation for human culture and civilization. Environmental apathy might even be said to reflect a lack of concern about the other problems faced by humanity. We may someday solve the major health, economic, and social problems that confront us. More likely, we will solve some, live with others, and face new problems as time goes by. In any event, we need the underpinnings of a strong natural environment. Without a healthy environment, the solutions we achieve will make no difference.

The High Cost of Environmental Degradation

Today's environment has already become far more unhealthy than we want it to be. In 1993 infectious diseases killed 16.5 million people, which represents one-third of all deaths worldwide. The number of people dying from infectious diseases exceeds those dying from cancer and heart disease combined. Such unpleasant mortality statistics for diseases such as ebola virus, malaria, and schistosomiasis are on the increase, according to a report by the Worldwatch Institute. One of the reasons for the increase is "severe environmental degradation." Diseases are costly, not only financially but also personally, in terms of physical suffering and loss of family members. Maintaining healthy environmental conditions makes sound economic sense for employees who are sick as well as for employers who pay for lost productivity.

In a report by Anne E. Platt titled *Infecting Ourselves: How Environmental and Social Disruptions Trigger Disease,* a strong connection is made between environmental conditions, economic development, and certain diseases. Those who battle attempts to maintain healthy natural ecosystems sometimes use economic concerns to justify environmentally detrimental development. Yet numerous ecological principles support the conclusion that human diseases increase in response to environmental abuse. Almost anyone would have difficulty asserting that pestilences, plagues, and epidemics are economically beneficial to humankind.

Platt's statement that "both the complexity and the diversity of natural ecosystems serve to keep infectious organisms and disease vectors in check" is significant. Most infectious diseases are caused by bacteria and viruses. A trait of some of these microbes is that they require other organisms, known as vectors, to infect us. Under normal, nondisrupted environmental conditions undergoing natural processes, the vectors that transmit such microbes to people, as well as to other animals and to plants, do not normally increase to levels hazardous to our health.

An example of how this relationship works is given in the Worldwatch publication. Lyme disease, unknown two decades ago in the United States, "now infects 13,000 Americans a year." Although a bacterium causes the ailment, the microbes are transmitted to people by ticks. "With the decline of wolves and other predatory species that cull herds, deer have multiplied, as have rodents, providing a moveable feast for ticks." This is indeed a simplified explanation for why Lyme disease has become more prevalent, or more apparent, than before. However, the principle is sound. When we decrease biodiversity of our natural ecosystems, the vacuums are filled by increases in certain species. What will happen in a particular habitat or situation is usually unpredictable, but when one species is allowed to increase in numbers because other species that normally control it are eliminated, the consequences for human beings are often detrimental.

Environmental degradation can be indicted as the cause of major assaults on human populations by malaria and yellow fever, once in the United States and today in many tropical countries. Both diseases are transmitted by mosquitoes that thrive in areas of dams, irrigation canals, and drained swamps. Mosquitoes that carry infectious microbes are often less abundant in natural swamp or river systems than in areas where improper environmental development has occurred to create unnatural breeding sites. The natural diversity of organisms keeps the number of mosquitoes in check. Much of the hidden biodiversity is eliminated with environmental disruption.

According to the Worldwatch report, human-caused environmental conditions are responsible for the alarming emergence of some infectious diseases. One reason for the strong connection between disrupted environments and diseases is that microbes multiply at rapid rates and thus can adjust to changing conditions faster than larger, more complex species. Their ability to adapt and respond quickly is one key to their success during massive environmental disruption. Disease-causing microbes are opportunists of the highest order.

Among the diseases included in the report are malaria, hemorrhagic fever, rabies, river blindness, and dengue (breakbone) fever. The environmental threats listed are deforestation, improper agricultural and irrigation techniques, dam and road building, and poor sanitation. Some of these contributing factors (roads, dams, irrigation canals) are perceived as "economic progress" by some people, but we must ask ourselves if the monetary profit can in any way justify the increases in a variety of diseases. Surely most people would agree that the answer is no.

The issue is whether we will acquiesce in the continued destruction of natural ecosystems for short-term economic gains for a few individuals. It is an important issue. For some the answer is a matter of life and death; for others it is a matter of whether they make more money than they have time to spend it. We must follow the example of the other creatures with which we share a large but finite amount of land and water. We must follow the same basic environmental guidelines that other animals follow. Destroying our natural environment is not a suitable long-term guideline.

Natural Laws Were Not Made to Be Broken

Throughout the world, rules regulating human conduct are constantly changing. Congress passes laws. The president vetoes bills. Courts hand down rulings. Rulings are overturned by higher courts. Heads of state sign peace agreements; others declare war. The rules change and change again. But the rules governing the plants and animals in the Okeefenokee Swamp, the Mojave Desert, and the Appalachian Mountains remain the same. The instinctive behavior of denizens of the swamps, deserts, mountains, and forests does not change in response to laws passed by Congress, court rulings, peace accords, or declarations of war.

Why this apparent distinction between human beings and other animals? Why do human societies continually change the rules, locally, nationally, and internationally, whereas natural communities of animals seem always to follow the same set of rules?

As far as nature is concerned, the answer is simple: There is no distinction between humans and other species. The rules that govern the natural world apply to people. The fact that some human beings do not acknowledge these natural rules, indeed *natural laws,* does not make them less applicable.

One of these natural laws applies to overpopulation, a topic of the 1994 meetings in Cairo to discuss the world population problem, a topic that will increase in intensity, never to go away as long as we have more humans each day than we did the day before. Under natural conditions no animal species is permitted to increase its population size beyond the resources needed to sustain it in a region. Nations, religious organizations, and a notable politician or two who do not acknowledge this rule for human beings did not participate constructively in Cairo and did not contribute to a better society for any nation or the world.

In the natural world, feedback controls operate to limit population sizes in various ways. Among animal species, if population sizes increase to levels that cannot be sustained with available food, females of some species respond by producing fewer young, thus lowering the population size in the next generation. Some birds actually lay fewer eggs during the nesting season in years when the insect supply is low. For a species that provides parental care, like birds or humans, the individual female can better ensure adequate nourishment for all babies if the number is limited. Concern for the welfare of her offspring being of utmost importance, she does not produce more than she can care for. Birds and other wildlife species display the ultimate "family values."

Another natural control is that animals living in an overpopulated situation with a low food supply become weakened and therefore more susceptible to predation and disease. Thus, the population size is reduced directly, through elimination of individuals, and indirectly, through a reduction in birth rate.

How do people avoid these feedback control mechanisms that effectively manage population sizes of all other animals? They do not. Not indefinitely. Evidence of this fact is painfully apparent: millions of human beings have starved, literally to death, in Africa and Asia during the last decade, and millions are starving today. It is a simple rule of nature: any animal will starve when more individuals of the species exist in a region than there is food for them to consume.

One distinction between our species and other animals in the natural world is a reliance on technology. A misconception of many people, often well-meaning people, relates to the effectiveness of technology, but technology cannot change the fact that humans have already reached overpopulation levels in some regions, to a point that natural controls are beginning to operate—starvation, disease, fighting. Perhaps those who do not acknowledge the natural rules that govern us should reconsider their responsibilities. Nature does not change the rules just because we ignore them. We are paying the consequences of breaking the rules, and the consequences can do nothing but become even more severe in the future. Politicians can continue to ignore the issue of population control, but the problem will not disappear with neglect.

The Wildlife Unemployment Rate Is on the Rise

Although most politicians eschew the topic of overpopulation, political rhetoric does include frequent references to "creating and eliminating jobs." Politicians also have a thing or two to say about "taxes" (although officeholders sometimes change the word to make it more palatable to the voting public). Human societies devote substantial time and energy to finding work and paying taxes. Are these activities part of the natural world? Do wild animals have jobs? Do they pay taxes?

Every raccoon in North America spends part of each day looking for food—that's a raccoon's job. Beavers are employed every day and night in repairing dams or cutting down trees to build lodges. The social insects, such as army ants, are a paragon of an organized labor union, although they never have strikes or contract negotiations. Each army ant has a job and is on duty every day. Some operate as defenders of the colony; some work to carry food to the nest; some serve as nursemaids to the queen. Yes, all wild animals have jobs.

The first order of business with any job is to acquire the necessities of food and shelter. Health care, defense, and transportation are also requisites of life—for human beings and for other animals. In human societies debates about how much of our personal income should go to pay taxes for these essentials are in the news every day. Clearly, people have not developed a universally acceptable formula for acquiring and distributing these necessities. How do other animals handle the situation?

To make an effective comparison between our species and

other animals, we must first identify a commodity equivalent to money. In an ecological sense, the answer is simple. Energy is comparable to money. A consistent supply of energy is essential for all animals, including human beings. To grow, to reproduce, simply to survive requires energy, which animals (including people) acquire in their food. A field of science known as ecological energetics attempts to measure how different species partition energy among the essential requirements for life.

To acquire energy, an animal in the wild must keep working. If an animal loses its job, it loses the ability to acquire energy, and it dies. Human jobs are regarded as "lost" if a business shuts down or reduces production. The same thing can happen to animals. For example, the endangered wood stork has declined steadily in numbers for several decades due in great part to a case of severe "unemployment." A wood stork's job is to find food in shallow wetlands where fish and tadpoles are concentrated. They have been trained for their job over evolutionary time, for millennia, and those that are extant spend every day pursuing this work. When people destroy wetlands by draining swamps or small aquatic sites, by replacing them with malls, parking lots, or housing developments, the wood storks no longer have a job. They are incapable of obtaining food except in shallow wetlands.

The unemployment problem we have created for wood storks is only one example of the economic toll we are taking on wildlife. Thousands of wetland species have their work places abolished each year. The same is true in forest habitats, deserts, even the oceans. When we kill the coral reefs by polluting a coastal habitat, we eliminate prey species that other species depend on for their food and energy.

Human jobs are, obviously, a matter of importance, but so

are jobs of wildlife. We put other species out of work—permanently—far more often than we do people. Natural systems provide services essential to our own existence. We should not put them out of business.

Do animals pay taxes? Absolutely! People tax animals as well as plants every day. We eat them, we wear them, we make medicines out of them. We use them for trinkets. The tax burden on them, paid with their lives, is heavy. We add to it further by needlessly destroying plants and animals in pursuit of our own goals. Our present approach constitutes overtaxation of our native wildlife, and the national wildlife deficit increases annually. In the long run, our most important tax base is natural habitats and the wildlife that dwell there. We cannot continue to exploit them in unsustainable ways without eventually eliminating that tax base.

Let's Hear It for the Herpetofauna

Serving as a solid example of some of the problems are reptiles and amphibians (collectively known as herpetofauna), which qualify as among the most maligned of our wildlife. They have taken a terrible beating from the U.S. public. Snakes are still killed by the uninformed; basking turtles are still used for target practice; salamanders are still used for fish bait. In addition, millions of reptiles and amphibians die each year on public highways, although most drivers are unaware of their contribution to the roadkill tally.

Public attitudes about wildlife protection, however, are changing, and herpetofauna are a barometer of this change. More and more people accept native reptiles and amphibians as a valuable part of our natural heritage deserving full protection. Educated people appreciate the key environmental role herpetofauna play as critical links in the food chain of which all organisms are a part. The environmentally informed realize that these animals are as important to natural ecosystems as those species perceived as directly useful to human beings. And many view intentional destruction of these animals as inexcusable.

Therefore, it seems worthwhile to identify other major threats faced by today's herpetofauna. Their wanton killing may be distasteful, but two other threats are even more serious. These environmental acts can eliminate entire populations of a species and ultimately drive the species to extinction. Today's greatest threat to most herpetofauna, indeed to all wildlife, is habitat destruction. Wetlands degradation, destructive lumbering, atmospheric and stream pollution, and the overuse and abuse of herbicides and pesticides continue to threaten the existence of countless wildlife species. A second form of destruction is related to the pet trade—overcollecting, whether legal or not. Understand that we have no problem with keeping reptiles and amphibians as pets. In fact, if a child or adult wants to catch and keep a native species of turtle, frog, or harmless snake as a pet, we encourage it if they do it right. The majority of today's professional herpetologists probably started their careers, as children, with a pet snake or turtle. Enormous numbers of herpetofauna perish each year through ignorance, carelessness, and indifference, but keeping them as pets does not have to be detrimental to the environment or harmful to the pet. However, certain guidelines should be followed. Obviously, any pet should be given proper care and attention, and the animal should be obtained in a legitimate manner. Also, animals returned to the wild should be released in healthy condition at the site of original capture.

How do we control commercial impacts on this component of our native wildlife? Most nongame species have no official protection. Thus, much of the overexploitation of our natural resources, whether through destruction of natural habitats or wholesale removal of the animals themselves, is morally and ethically wrong. But it is technically not illegal. State wildlife departments should limit the removal of an excessive number (which would vary among species) of any native, nongame species. Removal limits would apply until commercial wildlife dealers provide evidence that natural populations could sustain themselves at the higher removal levels. The process works this way for game species; the same formula should be used for other wildlife.

A similar approach could be used for commercial development projects that alter natural habitats. We should define the default option as being that native wildlife will be detrimentally affected by any environmental modification, unless evidence to the contrary can be given. Responsibility then lies with the developer to document that the environmental impacts would not result in a net loss of wildlife.

Pay Now or Pay Later, with Interest

These suggestions are in vivid contrast to the current system in which evidence of environmental harm must be given before development in natural habitats or overharvesting of nongame wildlife can be curtailed. Such a plan as we propose would call for a major environmental attitude adjustment among many citizens, especially commercial collectors and land developers, but it is an adjustment that needs to be made. Now. Let's reverse the formula.

Change in a lot of our economic-environmental formulas is needed. We have a proposal that should appeal not only to

Congress but to all the American people, a plan that would decelerate the loss of natural wildlife habitat, increase job opportunities where jobs are most needed, rebuild our cities, and provide tax incentives for ecologically sensible development.

We have heard a lot of talk about enterprise zones designed to entice developers into restoring inner cities. Nevertheless, large urban areas—human-supporting ecosystems—continue to deteriorate, becoming less and less useful to our society. At the same time, undeveloped areas—natural ecosystems—continue to be converted into human habitats at an unknown ecological cost to our society. What's wrong with this picture?

Much of our land today is lost to development because we, as a nation, condone practices whereby natural habitats that provide homes, food, and other biological requirements for wildlife are not recognized as having economic value. We can change this view overnight and begin to profit the next day.

Let's require developers to pay a fee equal to the value of the natural environment they want to develop. Areas with high natural ecovalue would then be prohibitively expensive to develop, and places with low ecovalue would be cheaper to buy. Regions with zero or negative value (such as decaying urban areas) could even offer financial incentives for development—paid for by fees collected from other development projects.

How would the monetary value of a natural habitat be determined? First, the ecological value of land holdings would have to be calculated. Heritage Trust programs, now firmly established and effective in many states, are constantly involved in this process, examining property in terms of how representative it is of the natural habitat of a region and whether it harbors native plants or animals that are ecologically important.

Whether a piece of property reflects the expected natural biodiversity of a region would also enter into the formula.

No matter how the ultimate assessment is made, it will be no more difficult than assigning value to commercial property. Sure, conflicts will arise in disagreements and different interpretations of how valuable a particular habitat or environment really is. But the formula for ascertaining economic value of natural ecosystems would be no more complex than any situation in which people create a yardstick for determining quality.

The ecovalue plan would also encourage and support programs in universities to train students in wildlife biology and basic ecology. They could learn how to assess the value of land and waters. The U.S. Fish and Wildlife Service, along with a variety of state conservation departments, already has this capability to some extent. Such programs, however, are woefully understaffed.

Once a plan to assign ecovalue to natural habitats is established, developers and landowners might choose to turn a natural ecosystem into something else, but, in most cases, areas that make the most ecological sense to develop would also make the most economic sense. Financial incentives would encourage developers to use land whose natural ecology we have already ruined, before developing others.

Who would oppose such a plan? Developers who trade in quick profits from low-cost purchases of natural lands with high ecological value might not be enthusiastic about having to purchase a costly building permit. People who make a living installing power lines or constructing highways might be affected by the increased costs in gaining rights of way through wildlife habitats. On the other hand, they might find work restoring and upgrading previously developed areas. Landowners who have not already severely disrupted the native wildlife might be disturbed because their land would now be less attractive for development, but the ecovalue plan could provide a decrease in ad valorem taxes for these landowners. That is, as the ecological value of the land increases, the tax per acre would decrease.

Would consumers object to establishing a plan to place monetary value on natural ecosystems? Probably not. Although increased development costs are always handed down to the consumer, the cost of renovating older, already-developed areas would be decreased. Developers and consumers should benefit from this plan. Furthermore, we would know we were developing or paying for the least ecologically expensive way to grow. We can pay now, or we—and our children—can all pay later. The choice is up to us.

Index